To Dr. von Mehren with

best wishes

Daniel Kasella

Magic
Cancer
Bullet

Magic Cancer Bullet

How a Tiny Orange Pill Is Rewriting Medical History

DANIEL VASELLA, M.D.

with Robert Slater

HarperBusiness
An Imprint of HarperCollins*Publishers*

HarperCollins books may be purchased for educational, business, or sales promotional use. For information, please write to: Special Markets Department, HarperCollins Publishers Inc., 10 East 53rd Street, New York, New York 10022.

Designed by Nancy Singer Olaguera

Library of Congress Cataloging-in-Publication Data:
Vasella, Daniel
    Magic cancer bullet : how a tiny orange pill is rewriting medical history / by Daniel Vasella, with Robert Slater.
        p. cm.
    Includes index.
    ISBN 0-06-001030-4 (alk. paper)
        1. Imatinib. I. Slater, Robert, 1943- II. Title

RC271.I43 V375   2003
616.99'4061—dc21

                                                          2002038769

03 04 05 06 07 ❖/QW 10 9 8 7 6 5 4 3 2 1

The story of Gleevec is the story of the discovery of a breakthrough cancer drug told through the fears and hopes of patients and with the voices of scientists, doctors, and many other individuals who contributed to making this dream a reality. To all of them and to my colleagues at Novartis who worked tirelessly to bring this medicine to market rapidly, I would like to express my gratitude. It is, finally, to all of the patients who have been helped by Gleevec that I would like to dedicate this book. They were our inspiration throughout in the story that unfolds on the following pages.

—Daniel Vasella, M.D., CEO, Novartis

Contents

The Dreaded Word

Patient Tales

Etched in Stone

It was December 1995. Judy Orem was 51 years old, and feeling fine.

It had been two years since her last check-up. She figured it would not hurt if she had some blood tests.

When the phone rang the next day, she wondered who might be calling; she had forgotten the tests altogether.

On the other end was her doctor. She spoke to her in a tone that she would never forgot. Her voice was solemn.

She informed her that she was sending her to a specialist for a bone marrow biopsy the very next day.

Before Judy could ask why, she said that she believed she had chronic myeloid leukemia (CML).

The only word she heard was "leukemia."

Her mind raced. Leukemia. Cancer. Terminal disease. Death.

In the coming days and weeks, she grew familiar with medical language and measurements that had been totally foreign to her.

She learned about white blood cells and that an increased number of them tended to confirm a diagnosis of CML. She learned that a normal white cell count was between 4,000 and 10,000 per cubic millimeter. Hers stood at 68,000.

During one visit, her doctor told Judy she had three to five years to live. It all seemed so etched in stone, so final.

She thought of her life, which revolved around children, a life of scouts, soccer, youth groups, and volunteer work.

She was born in Portland, Oregon on March 7, 1944 and taught 7th and 8th grade math and science at an all-girls school there. Then, she chose to stay at home as her two children were growing up.

She loved her life. No matter how gloomy the doctors were, Judy was not willing to give up. Yet, the only cure for CML victims had a lot of catches.

The one cure was a bone marrow transplant, but the procedure is highly dangerous: 70 percent of the patients who undergo this procedure do not survive beyond a year.

A Spark of Hope?

But if there was even a spark of hope, she was willing to pursue the spark. And so she traveled to the Stanford University Research Hospital in Palo Alto, California to look into a bone marrow transplant.

She was told the chilling news that she had no more than a 50 percent chance of surviving the actual procedure. Increasing the odds against her was the fact that she had no siblings who might have been suitable donors.

She opted for the three to five years that the doctors had spoken about. She was not willing to gamble on the high-stakes transplant option.

The doctors wanted to put her on interferon, the drug of choice for CML patients, if she chose not to do the transplant.

Because interferon reduces one's white cell count to tolerable levels, it does prolong the life of a CML patient, not dramatically, but long enough to make it the most popular therapy for that disease. Treatment with chemotherapy led to a survival rate of four years; interferon, six years.

For CML patients, interferon had another advantage over chemotherapy: While chemotherapy had no effect on decreasing the number of Philadelphia chromosome–carrying cells, the culprit in the cell that is the main marker for CML, interferon removed a large majority of these cells in 15 to 20 percent of CML patients. This so-called major cytogenetic response clearly led to longer survival rates.

Still, interferon was no cure for CML. Cancerous cells remain in the person's bone marrow and even with interferon treatments eventually cause death.

Judy remained on interferon for three years, never falling below the level of 60 percent Philadelphia chromosome positive. She had unpleasant side effects: a metallic taste, loss of weight, and low energy. In December 1997 she started taking one of the older chemotherapy drugs called Ara-C. She stayed on that drug for six months. Her short-term memory grew less sharp.

Three years had gone by since Judy had been diagnosed with CML. She was hoping that the interferon would give her a few more years—and then some. But she was not overly optimistic.

The Tiger Is Sleeping

July 1997. Thousands of miles to the east of Portland, Oregon, Marco Nese, a Swiss-born businessman from the town of Basel, had no particular aches or pains other than his usual migraines. A couple of months earlier his father found that his cancer had reached a very advanced stage. Marco's concern for his own health intensified. He decided to go for a check-up.

The doctors put Marco through a battery of tests. Then came the shocking news.

The tests showed that Marco's white cell count had climbed to 41,000. Marco was suffering from CML.

Marco was so young, only 33 years old. His doctor reluctantly broke the awful news to him that he probably would not live for more than five years.

Suddenly, life seemed terribly short.

Marco was born on November 7, 1963. He has roots in both the academic and business worlds, receiving his doctorate in sociology in 1992 and then embarking upon a career in marketing.

Unfortunately, being ill was not new to Marco. From the time he was nine years old he suffered migraines. He was constantly taking anti-migraine drugs. Then, when he was 22 years old, he was diagnosed with a rheumatic disease and he took more pills.

Marco's options for therapy to treat his CML were limited. Nothing promised to be more than temporary. He had no siblings, eliminating any hope of finding a family match for a bone marrow transplant. Marco searched for a donor, but to no avail.

A month after being diagnosed, he began taking interferon, conscious that it was no cure: "You still have the disease in you, but the tiger is sleeping. One day he will wake up. You take interferon to keep him asleep."

Marco's white cell count dropped to normal ranges but it was likely to be only a temporary improvement. Interferon has painful and debilitating side effects. The daily injections of the drug just about shut him down: He began sleeping 13 hours a day, constantly feeling tired; he went into massive depression.

Interferon might double his life expectancy—from five years to ten; but those ten years could hardly be called living.

At first, Marco continued working in marketing and sales for a Swiss business; but then, early in 1999, with the drug taking its toll, he cut back to part-time. He was not willing to give up on life completely. He began an MBA program in London at the Ashridge Management College.

But his health began to deteriorate further.

Early in 2000, he suffered through the first of three bouts of pneumonia that lasted into the summer. He continued taking interferon but its effects left him unable to work effectively. He eventually stopped work entirely.

He felt that his life was slowly slipping away.

How Much Longer Do I Have?

In November 1998, Darlene Vaughan's physician discovered a large abdominal mass. The CT scan indicated that Darlene had a benign tumor, Two weeks later, however, the results of tissue slides showed that she had Stage 3 leiomyosarcoma, a rare cancer that occurs about four times in a million, and is very resistant to traditional therapies including radiation, chemotherapies and surgery.

Darlene was born on October 3, 1942 in Pasadena, California. She attended California State University, Long Beach for a year but dropped out to begin a career in business and project management in the aerospace/electronics industry. She was married briefly in her 30s. Abandoning the corporate world, she moved to Pueblo, Colorado. She opened a "San Francisco" style coffee house/cafe in November 1994, living above her place of business, the upstairs of a historic building that had once been a brothel.

Making a living from the business was tough, but she persevered until 1997 when she became unwell.

One Step at a Time

At the time of her cancer diagnosis, Darlene's oncologists suggested radiation therapy. Darlene, however, wanted to seek another opinion.

On December 12, 1998, Darlene went to see cancer specialist, Dr. Lee S. Rosen, at the UCLA Bowyer Oncology Center in Los Angeles. She returned to UCLA in late January to begin seven weeks of radiation treatment.

She then traveled to Colorado to take time to recover from the side effects of the radiation. A CT scan of her pelvis in June 1999 revealed nothing worse. A CT scan in October 1999 showed a metastatic lesion on her liver. In November, a surgeon at UCLA began surgery to remove one lobe of her liver. During the operation, however, doctors discovered a number of small lesions too small to show up on the CT scan and left the liver intact.

Because of the presence of numerous legions, they decided that her disease required systemic treatment; i.e., chemotherapy or experimental drugs, as opposed to a localized invasive technique.

While recovering at home from the second surgery early in the year, she put herself on an intensive regimen of vitamins, supplements, exercise, meditating and journal writing, hoping to allay the growth of the tumors.

Darlene realized that she could not fight the cancer and run the cafe at the same time. She finally sold the business and the building.

In the spring of 2000, Darlene began taking an experimental antiangiogenesis drug called SU-6668. Two months later, a CT scan showed that the three largest tumors and now numerous other smaller tumor deposits had grown. In August she began taking a traditional chemotherapy drug called Dacarbazine, or DTIC. The drug caused significant side effects: her fever spiked to 105 degrees, she was nauseous, had hardly any appetite, and was very tired.

The next CT scan indicated that her tumors were growing; they were numerous and were getting quite large. By the fall of 2000 she began to fear the worst. She kept asking herself: How much longer do I have?

≡

What did Judy Orem, Marco Nese, and Darlene Vaughan have in common?

They had been leading fairly normal lives before disaster struck.

When they were diagnosed with cancer, they all went into shock and depression.

Each of them knew that the odds against beating the disease were slim indeed. Each of them wanted to fight as hard as possible to hang on to life.

But none of them thought they had much more time on earth.

All three prayed for a miracle.

Introduction

We felt what we were doing was right. But deep down we sometimes had doubts too.

Whatever others might say, we were convinced of our mission. What could be more important than rescuing a patient from the brink of death?

Every doctor—and I include myself—when faced with a patient in a desperate situation, wishes for a therapy that will cure that patient, a drug called a "magic bullet."

It is of course the scientists who search for such drugs in their laboratories, and, if they succeed, those magic bullets will emerge.

Though I had little taste for the laboratory, I always felt deeply for patients; I wanted to assist them, to support the healing process; at best to save their lives, or at least to prolong and improve their lives. So I became a physician and later entered the pharmaceutical business.

Altering Medical History

Never in my wildest dreams did I think I would run a pharmaceutical company, let alone one that would discover a breakthrough cancer drug. And yet, thanks to Novartis's discovery and development of this wonderful little orange capsule that we have named Gleevec and that is the subject of this book, I have had the good fortune to have a hand in altering medical history.

Since 1996, I have been CEO of Novartis AG, a Swiss company that owns one of the largest pharmaceutical companies globally. Our main mission is to discover and develop drug therapies that will help prolong, improve, or even save lives of patients.

As the leader of a major multinational pharmaceutical company, it is my duty to help create a fertile environment for drug discovery and development by choosing and supporting the right scientists: those who take calculated risks, who are dedicated to our mission; they should also have aspirations and be superb professionals. Usually I begin to oversee the actual drug making process when we start development and commit significant resources.

We at Novartis are engaged in so many innovative efforts, we are trying to conquer so many new scientific worlds, that it would be impossible for our scientists to keep me in the loop on every compound that is being discovered. Only one of every ten thousand molecules eventually makes it; all others are dropped at some stage in the development process because of insufficient efficacy or side effects. So many experiments, and so many failures, it is all so amorphous. One of our scientists may come upon me by chance and indicate that this particular compound looks promising, or that another one seems hopeful. But ultimately I must wait for them to deliver results indicating that a compound truly has potential.

In the case of our breakthrough cancer drug, Gleevec, I remember that first moment when the evidence arrived that it worked in humans.

The first results from Phase I patient trials for the drug's effect on CML patients came in April 1999. Normally we would never see such results, for Phase I trials rarely uncover evidence that the drug is working. At this early stage, all the clinicians usually want to do is learn how a drug is absorbed, metabolized, and excreted. Those studies are often performed on healthy volunteers. Only with cancer does one normally start with actual cancer patients because cancer drugs are generally potent with severe side effects.

At first, I had trouble believing the data.

The numbers told of 31 Phase I patients, all of whom had taken the drug at least at 300 milligram doses; everyone experiencing their white blood cells being reduced to normal levels.

Never before had any drug proven that effective on CML patients. Not radiation, not chemotherapy, not interferon, not Ara-C, not even the only known cure for the disease, a bone marrow transplant, which results in the death of many patients.

All of these therapies had made some inroads into the disease; but none showed such a high percentage of success and so early in the treatment.

With such spectacular results in hand, the next steps were obvious: Direct large amounts of capital and human resources to developing this drug. Rush the drug to market.

How could we act otherwise?

After all, who could forget that the lives of patients were at stake? What we really cared about were all the patients who were waiting for a better, safer and more effective drug.

Who could forget that too many of our scientists knew only failure throughout their long careers with us? I of course knew of all those moments of hope and frustration our scientists had to deal with every day. And now that our scientists had unleashed what appeared to be a breakthrough cancer drug, how could we not go full steam ahead in getting this drug to market?

Choices to Ponder

But, in fact, the next steps were not at all obvious. The pharmaceutical management team and I had some agonizing choices to ponder given that the number of CML victims in America was relatively small compared to the victims of other forms of cancer and other diseases. Some 6,650 Americans over the age of 10 are diagnosed with CML each year; worldwide, the figure is 1.3 per 100,000 a year. And yet, nearly 200,000 cases of prostate cancer, and 195,000 cases of breast cancer were expected for 2001 in the US.

Ultimately, we are a business, and business decisions are often based on some form of statistical analysis and the chance to make a profit.

If it were purely a personal choice, we would ignore the questions that business professionals must ask. We would avoid asking how many patients suffer from CML. We would not ask whether we could supply this drug without charging a high price. We would not ask what would be the costs to the other parts of our business, if we threw large amounts of capital and human resources into the development and manufacturing of this drug.

We would skip all those questions. But we cannot for, as I said, we are a business.

Confronted with those unprecedented medical results on that April day, we at Novartis were facing difficult decisions. Even if we produced a breakthrough cancer drug for CML, we would affect only a tiny portion of cancer patients. In addition, we faced the uncertainty that surrounded the Phase I results. How could we know for sure that the drug would prove successful?

Believe me, I have no magic formula in making such decisions. I could only fall back on my experience; I could listen to my colleagues and try to sense how logical and enthusiastic they were about the drug. Ultimately, there is some intuition that guides our decision-making at Novartis.

Sometimes that intuition—and I admit this—leads in the wrong direction; but fortunately, in most cases, in trying to choose winners for our drug pipeline, we have a strong sense of when to push forward, and when to halt a project.

But certainly not always.

As word of Gleevec's special properties began to surface, we had to make a tough decision: whether to break out of our usual drug

Magic Cancer Bullet

development pattern by steering a large amount of resources—both capital and human—at the drug far earlier than normal. We would be heightening the financial risk and perhaps creating huge "opportunity costs" by diverting resources that would have been earmarked for other drugs.

The data, my intuition, and thoughts about the patients, led me in only one direction.

≋

On that day in April, I immediately phoned our head of development at Novartis, Dr. Jörg Reinhardt.

I told him that I was looking at the Phase I results for STI571 (the name for Gleevec during its development).

"This can't be," I said to him, assuming that something was wrong, that somewhere there was an error.

Finding errors in such results was uncommon but not impossible. Sometimes, one feels a surge of excitement, only to find that a lab error has occurred. Part of me wanted to jump for joy; but an inner voice warned, "Don't get excited. First double-check."

As if reading my mind, Jörg observed, "We *did* double-check. That's how it is. The data really looks pretty good."

His voice sounded firm, confident, and excited. For Jörg to get excited is a definite sign that something is up. He is normally about the most sober and balanced fellow imaginable.

As I spoke with Reinhardt, it became clear to me what we should do.

"We have to speed up the trials," I asserted.

Let's Not Count Pennies

We decided then and there that the number of patients who might benefit from such a drug should not concern us. As a management team, we have a duty to put a product on the market when there is a fair chance that it will change the practice of medicine—even if it only helps a small number of patients. In that sense, we certainly do not function like many other corporations.

I knew what we had to do, that we must not count pennies and must disregard costs. We would find the money. "I don't care what it costs," I told our head of global technical operations, Andreas Rummelt. "Money doesn't matter. Let's just do it."

Once I had made up my mind to press forward with the drug, Jörg tried to bring me back to earth. "We don't have enough drug substance."

I knew I sounded harsh, but I could not help myself: "Just make it."

Jörg wanted to make me understand: "You know Dan, it's not so simple."

I knew that; but even if on that day, it all seemed an impossible task, I was completely confident that my colleagues would find a way to accelerate the process—to produce enough drug substance so we could speed up and expand the patient trials.

Faced with the extraordinary data, I sensed that Novartis was bound to come under heavy pressures—from the media who would monitor our every move, delving into every aspect of the drug, its pros as well as its cons; from physicians who would hope to get the drug for patient trials; from individual patients who would want the drug; and from patient groups who would lobby us to get rapid access to it.

Let's face it: Patients have profoundly changed their behavior.

These are not the patients of earlier days who felt ashamed to talk in public about their illnesses; who relied entirely on their doctors for wisdom, knowledge, advice; who, once they were struck with a serious disease, were kept uninformed about their illness.

Today's patients are a demanding lot. They assert a right to decide and a right to the best treatment. They hunger for information about their disease. They have taken to the Internet with a vengeance, and because of the resources available on-line, they are smarter than ever, discovering drugs at a very early stage of development; alerting their physicians to new drugs and therapies.

This Is a Time Bomb

I had every reason to believe that this Internet generation of patients had the potential of making life difficult for a pharmaceutical company that did not live up to their expectations.

After all, the earlier example of AIDS patients transformed into activists against pharmaceutical enterprises gave all of us at Novartis pause. We had to display our concern for CML patients; we had to show them that Novartis was doing all that it could to speed production of STI571. I told my senior colleagues: "This is a time bomb, and if we don't handle it well and act fair, it will blow up in our faces."

In speeding up the production process of the drug, we increased the costs of failure significantly. Our plan was to invest heavily in production up front without waiting for the results from further patient trials. That was an extraordinary decision, but we made it, knowing that we were enhancing the risk enormously.

Probably some of my colleagues thought that I was completely nuts. But I did not care. It was clear to me what the right thing to do was. If we had followed the established, sequential ways, it was safe to say that over the next year, thousands of patients would not get a drug that might have alleviated their suffering, prolonged, or even saved their lives.

This was unacceptable.

Finding delay unacceptable, we acted aggressively to get this tiny orange pill into the hands of patients as quickly as possible. This book tells the story of how we did that. I never thought I would write a book about a pill.

But here I am, spending hours with my writer-colleague, Robert Slater, trying to decide how to tell the dramatic story of Gleevec, working hard to convey the medical significance of this drug as well as the human emotion displayed by the real stars of this drama—the patients; making sure to tell the business side of the story as well.

The truth is: I never thought of myself as an author. A doctor? Yes. The Chairman and CEO of a major pharmaceutical company? No.

As a business leader, I have watched and presided over one of the great business stories of our era—the arrival of Gleevec to market in record time for a cancer drug. Behind that larger tale are all sorts of stories, occurring in various settings—from the laboratory to the patient clinic; from the boardroom to production facilities and patients' homes.

From my unique vantage point I have watched these stories unfold. Truth be told, I had nothing to do with the science that breathed life into Gleevec. Yet, I know many of the scientists and have stayed in close touch with them throughout Gleevec's birth pangs.

Within what is possible, I have done all that I could to ensure that Gleevec is available to as many patients as possible in a timely fashion.

Having the privilege of overseeing so many of these elements, I wanted to write a book about how long, complex, thorny, and risky it is to discover and develop a breakthrough drug.

I wanted to explain how difficult—yet in a paradoxical way, how easy—it was to decide on pushing Gleevec to market. I wanted to take readers into the inside of a large pharmaceutical company—in this case our own Novartis—to give a sense of how challenging it is for us to produce a drug like Gleevec, given the pressures on us. The pressures come on two critical fronts, patents and pricing. We must be able to protect our patents for drugs; otherwise there will be no incentive for our scientists to be innovative and share their findings. And we must structure our prices high enough to assure a return on investment sufficient to support ongoing research and development.

Through this book we have the chance to explain that unless we can cope with these challenges, there can be severe consequences for future investments that form the financial backbone for our research and development effort. Without those continued investments, without the framework to survive economically, there will be no more drugs like Gleevec.

Through the telling of the story of Gleevec, we will give you, the reader, an understanding of the complex machinery of drug discovery and development; and the exposure a company risks on patents and pricing.

Unique Vantage Point

I bring a unique perspective to the telling of this story: I have been privileged to see the story unfold in all of its aspects. It is because of that unique vantage point that I have decided to write this book. I sense a rare opportunity to tell the complete story.

After all, not many business leaders get the chance to manage the development and production of a breakthrough drug. Not many business leaders are likely to face the conflicting pressures that we confronted in making sure that Gleevec got to the right hands speedily.

Many business leaders may well have the experience of managing some exciting, new product. Many of the issues associated with managing a product like Gleevec and managing some other exciting, new product—whatever it is—appear to be the same.

How do you manage the product?

How do you cope with the pressures?

How does the CEO satisfy the various groups competing for time and attention and making contradictory demands?

In this book, I want to talk about the marvelous story behind the birth of Gleevec—the tale of great minds working toward the noblest goal, saving lives.

I also want to talk about the way a major international corporation (Novartis) "managed" that birth through the many complexities that arose.

We employed in part what I call "innovation management," which is appropriate when bringing a breakthrough product to market; it requires a host of skills, including the willingness to take great risks, an emphasis on speed and quality, and an ability to "think outside of the box."

We also employed "success management," taking all the hard decisions required when a product is a major success. An important part of this kind of management is dealing with the question: Who deserves credit for the discovery of the drug? We felt it important to let the world know how we parceled out credit for the discovery and development of the drug.

≋

The story of Gleevec of course is in fact the aggregation of many smaller stories.

First and foremost, there are the patient stories, as Gleevec rescued their lives, gave them new hope, and for some at the brink of death gave them a new start.

Patients played an important role in spreading the word about Gleevec, relying on the Internet, forming patient support groups, exchanging information, and lobbying Novartis aggressively.

You will meet these patients and read their stories. You will also read about the activities of patients who campaigned intensely to increase the supply of the drug at a very critical time.

You will also read the stories of the Gleevec scientists. While building on the work of 40 years of scientific research, these scientists had the foresight and imagination to theorize but even more importantly, to put into practice these theories and to give the world a completely new way of treating cancer. We will give you a glimpse into the laboratories and, without getting overly technical, explain the challenges that the scientists faced.

Next we will tell the stories of Novartis' own development and production teams who, sensing that lives were at stake, did the seemingly impossible by bringing Gleevec to market in record time for a cancer drug. Team members will recount how obsessed they became with getting Gleevec to patients, as rank-and-file employees evolved into crusaders.

Finally, there is my part in this story.

To be candid, though others have encouraged me to personalize this story as much as possible, I find it difficult and embarrassing. So many people took part in the discovery and development of Gleevec that often I felt but a small cog in a much larger wheel. The fact is that I think we have a wonderful story to tell, an emotional one, a dramatic one, and I have tried hard not to let my own thoughts and opinions get in the way of telling the story.

Molecules and Pathways

It's in Our Genes!

Why the excitement over Gleevec?

To begin with, for so long, cancer has been the equivalent of a death sentence. While much progress has been made in combating some forms of cancer, a death sentence still hangs over many, many cancer victims.

Against this dark landscape comes Gleevec.

No other cancer drug has exhibited so much success so quickly with tolerable side effects.

And for the very first time, cancer patients—specifically, patients passing through the early stages of CML—have eliminated or at least slowed down the cancer within them by simply taking a few pills each day.

This means no injection is required, no radiation, no surgery. It also means the absence of the common side effects in chemotherapy: vomiting, profuse diarrhea, infections, and bouts of massive depression.

The capsule allows most patients to lead a normal life.

To CML patients, the news about Gleevec has seemed too good to be true. Until now, their news has all too often been dark and tragic. Some 20 to 30 percent die within two years of diagnosis; and 25 percent die each year after that.

Designer Drug

The main reason for the excitement surrounding Gleevec is its innovative approach to cancer therapy. Indeed, if Gleevec did nothing more than help CML patients combat their disease, we would have much less cause for the enthusiasm this drug has produced.

What we have in Gleevec is nothing less than a scientific break-through that many medical experts say will revolutionize cancer therapy. The hope is that Gleevec represents a paradigm for other molecularly targeted drugs that will work on cancers with larger patient populations.

≣

The standard way to combat cancer has been to remove it surgically and/or use toxic drugs or radiation in order to destroy cancer cells.

These techniques also attack normal tissues and while curing some cancers, frequently leave patients overwhelmingly weak. These therapies have saved many lives, but their side effects can be so severe that patients even die from therapy. But these techniques were all that we had for many decades.

With Gleevec, the focus has shifted to a drug that specifically tar-gets a cancer-producing molecule, in effect shutting that molecule down, and thus keeping it from inducing cancer.

Gleevec offers the proof of a new and innovative concept, using genetic information and insights on molecular pathways. It repre-sents the first designer drug for cancer therapy, and one of the first examples of rational drug design arising from the new field of human genome studies.

Gleevec is "designed" to target only cancer cells, leaving normal, healthy cells intact. In that sense it is a great advance upon the most popular form of treatment for CML patients, interferon, a biotech drug.

To be sure, chemotherapy has helped thousands of patients. In a number of cancers, including Hodgkin's disease and some childhood leukemias, for example, over 90 percent of patients can be saved if the disease is discovered early enough.

Gleevec is the first of a new class of drugs called signal transduction inhibitors, so called because these "STIs" interfere with the pathways that signal the growth of cancerous cells. When an STI interrupts a signal transduction pathway, a cell stops dividing, halting the cancer.

With its new mechanism of action proving so successful, Gleevec has already ignited a good deal of research aiming at the development of more drugs using the same approach.

Of course, scientists had an advantage in trying to develop Gleevec: the early discovery that a genetic defect caused CML, i.e., that the disease was the outgrowth of a certain process operating via a molecular signal pathway. This encouraged scientists to seek out a specific compound that would combat the disease.

≋

Even before scientists get around to other Gleevec-like molecularly targeted drugs, they are eagerly investigating whether the drug will work on other cancers besides CML.

Scientists have noted that Gleevec inhibits the activity of three members of a family of enzymes—so-called kinases—that has over 100 members.

One family member was Bcr-Abl, the oncogene that resulted from the formation of the Philadelphia chromosome, the main marker for CML.

The other two family members are PDGF-R (Platelet-Derived Growth Factor) and c-Kit.

We at Novartis knew about PDGF-R and Bcr-Abl when we made STI571, but it was Dr. Brian Druker, working in his laboratory in Portland, Oregon, who showed that STI571 also affects c-Kit.

C-Kit is involved in GIST—Gastrointestinal Stromal Tumors—a rare tumor, which occurs in some 2,000 Americans a year. A GIST tumor can only be treated by surgery and unless eradicated, the disease is usually fatal.

C-Kit may also be involved in small cell lung cancer, accounting for as many as one-third of all lung cancers.

The PDGF-R enzyme is involved in a variety of cancers.

So, hope exists that Gleevec will work in other cancers where these enzymes play a role in cancer growth. Accordingly, Novartis has started patient trials in the generally deadly glioblastoma, a very aggressive brain cancer; some forms of prostate cancer; breast cancer in combination with other drugs; small cell lung cancer; and GIST.

The fight against cancer goes on and each new, effective drug like Gleevec is a small victory in a much larger, ongoing war.

In offering a whole new approach to cancer therapy, Gleevec opens a whole new category of cancer drugs. For CML patients, the drug may mean the difference between life and death; for cancer therapy, the drug offers proof that a new, promising field exists and has validity.

The excitement surrounding Gleevec is only natural considering the unshakeable, unvarnished truth that for still too many cancer victims the disease offers little other than a death sentence—or at least, a long period of suffering.

Until radiation came along in the 1920s and chemotherapy in the 1950s, cancer patients truly faced death sentences, except the minority who had successful surgery.

Some cancers could be treated effectively but overall most could not, especially those in advanced stages.

The harsh reality is that one out of every three people will be diagnosed with cancer.

Some 1,500 Americans die of the disease every day, more than 500,000 a year. Only heart disease kills more Americans.

At least 8 million people are diagnosed with cancer each year on a worldwide basis.

For centuries, physicians simply guessed at—usually incorrectly—the causes of cancer. Then in the early 1950s an event of great historic significance occurred, one that would have a profound effect on science and, more specifically, on the whole way we looked at cancer therapies.

Until this event, we knew precious little about how living things worked. To be sure, it was already known in the 19th century that living systems were designed as a collection of tiny, individual units called cells. Scientists deduced that a unit of heredity known as the gene was located on the chromosomes found in the cell's nucleus, separated from the rest of the cell. This then was the brain of the cell. It was not clear at all of what stuff the genes were made.

Experiments in the 1940s and early 1950s showed that the chemical transmitter of genetic information was DNA, or deoxyribonucleic acid.

The Secret of Life

The event occurred on February 28, 1953 when the English physicist Francis Crick walked into a pub in Cambridge, England, and announced that he and James Watson had found the secret of life. That very morning he and the American biochemist Watson had unraveled the structure of DNA. They suggested that genes were made up of four chemical units that were arranged along each of two complementary strands; the units form a code of instructions required for an organism's growth and reproduction.

DNA is thus the basic genetic material of the cell; it is the building block of all kinds of proteins. The DNA is located in the nucleus of the cell and is arranged in 46 chromosomes; hence, each human cell has 46 chromosomes, large-sized molecules, each one comprising numerous genes that encode the sequences that are the blueprints for proteins.

The Watson-Crick discovery would eventually encourage scientists to focus cancer research on molecular genetics. For some time— throughout the 1950s and 1960s—not much was happening in the search for a cure to cancer. Then two days before Christmas in 1971, President Richard Nixon signed the National Cancer Act, committing the United States to a war on cancer.

As scientists learned more and more about DNA, the way it replicates, and how to manipulate it, the field of molecular biology became increasingly important for cancer research.

We learned that we carried a set of genes that determined our individual height, eye color, and numerous other traits; and we learned as well that we also harbor genes that can give us cancer.

Until then, and this is incredibly significant, many regarded cancer as an outside intruder. Now, scientists were beginning to focus on the assumption that cancer came from within us. They came to accept that cancer was a genetic disease, in which some genes, when damaged, cause cells to grow out of control. Triggering cancer was a small set of genes, called oncogenes that control cell growth. An oncogene is a cancer-causing gene that results from the mutation of a normal gene.

While normal cells divide, replicate, and die off millions of times during the course of someone's life, during that process small mistakes occur that are built into the genome. Those mistakes can sit dormant for decades before the genes release a spurt of growth signals that order the cell to begin dividing and spreading very quickly.

What was truly significant about these discoveries was the hope they raised that a way could be found to treat or perhaps even to prevent cancer by altering the way these genes behave. The broad concept was that understanding what was broken would make it possible to fix what was broken, i.e., if we understood what was driving the growth of the cancer cell, we might be able to come up with a drug that would shut it down.

Crucial Questions

In this book, we tell the story of the effort to discover and develop one of those drugs.

The search for a cancer cure began to focus on those families of proteins inside the cell that controlled growth.

The scientists asked crucial questions:

What were the most promising parts of the cell to focus on?

And once targets in the cell were identified as promising, would it be possible to create drug compounds that could have an effect on cancer therapy?

One area of research that has had scientists excited is one of the first oncogenes identified in the early 1980s—the epidermal growth factor (EGF). The hope has been that monoclonal antibodies could be developed against EGF in order to impact a variety of cancers, including breast, lung, and bladder cancer. Antibodies, however, are expensive, and must be injected; moreover, they would be digested if given orally. Cheaper, small molecules were required.

Some scientists believed that it might be promising to search for small molecules that could block growth-factor receptors.

Unlike the antibodies, which react with the part of the receptor that projects out of the cell, the small-molecule inhibitors being researched act on the other end of the receptor molecule, the inside part that transmits the growth signals to the molecules of the internal signaling pathways.

For most of the cancer-causing growth-factor receptors, these internal segments are tyrosine kinases, which are enzymes that activate the signaling proteins by adding phosphate groups to them. A small compound lodged in the right place can often close an enzyme down.

But, as we shall see soon, developing these small molecules was easier said than done.

≋

Looking back, it seems a bit odd that I would ever be interested in such things as DNA, molecular genetics, or cancer research. It is odd too that I became a physician, for neither my mother nor father's side evinced much interest in the sciences or medicine. One of my uncles on my father's side came closest, practicing psychiatry. But he was no doctor. In fact, he was a priest! A real intellectual, he seemed to hold psychiatry dearer than religious faith, for when he died, I had a chance to visit his library where the number of books on medicine and psychiatry far exceeded those on religion. From my mother I learned we did have physicians in the family, but they were from the 18th and 19th centuries. During my childhood, I knew of no one in our family who practiced medicine—and medicine, accordingly, was far from my everyday thoughts at that time.

My parents, Ursula and Oskar, both grew up in Chur, the largest city in the mountainous canton of Grisons, largest of the Swiss cantons. My father's father, Pietro, with his wife Emilia, and the first of their nine children, Eduard, settled in Chur; they had walked over two mountain passes from the small valley of Poschiavo in Grisons in the south from where they emigrated as no jobs were available.

I never knew my grandparents on my father's side (my grandfather, a gunsmith, died in 1938, my grandmother in 1945); but from childhood stories, I have an image of hardworking, tough folk who labored from dawn to dusk in their store; encouraged their children to look after one another; and placed little value in nice clothing or a fine appearance. Of my father's siblings, only three did not obtain a university degree; education was valued far more than wealth.

My father was born in 1904, my mother twelve years later. Quite a learned man, my father spent his youth in various Swiss towns, attending Catholic high schools. Later, he pursued his studies in Berlin, Paris, and the Swiss towns of Bern and Fribourg. In 1928, at the relatively young age of 24, he became a professor of history at the University of Fribourg and remained in that post until his death, at the age of 62, in 1966 when I was 13 years old.

Born in 1916, my mother was one of five children (three sisters, two brothers). She attended elementary school in Chur, but when only eight, she together with her mother and siblings followed her father Joseph, and her older brother Joseph to Spain, in effect immigrating there in 1924.

The father and son made baby supplies in a small factory, but soon after the 1936 Spanish Civil War began, the family was forced, literally overnight, to seek shelter across the border. They wound up in Chur penniless. My parents met sometime thereafter and were married in 1942.

I was born in Fribourg, a small town between the French- and German-speaking part of Switzerland, on August 15, 1953. I was the last of the four children born to my parents. My brother Andrea was born in 1943; my sister Ursula, in 1944; and my other sister Silvia in 1946.

In 1960, my sister Ursula, then only 16, became seriously ill. Her battle for life became an inspiration to me.

I was just a child of seven at the time. Though I had experienced my own share of pain and suffering when I contracted meningitis and tuberculosis at age eight, no event—maybe excluding my father's death—saddened me more than her death a few days before her 19th birthday.

Her tragic illness and death had a lasting effect on me. More than any other event, it would inspire me in later life to help others as much as I could. It was the first time that I encountered the dreaded disease of cancer—and it was in a very close and personal way.

I recall Ursula showing me the burn marks on her skin from her radiation treatments. And I recall her losing her appetite completely toward the end. She refused to give up, finishing high school that summer before she died.

In her last days, she had lost all strength; she was down to skin and bones and had trouble breathing. Because her cancer had spread to the liver, she developed jaundice. My father had yielded his side of the bed to Ursula while mother remained on her side.

Ursula knew the end was nearing. On that afternoon in December 1963 when she finally succumbed to her illness, she told me: "Daniel, do well in school."

Suddenly, for the very first time in my life, I was forced to absorb what a person looks like in death. I still possess a sharp recollection of touching Ursula's once-warm hands, now cold. I could not get over that her skin did not move, that her face was motionless. I wondered what it was like to be dead. I was just a boy of 10, but I was growing up very fast.

I did not understand back then what *cancer* was; that came later, as an adult, as I began studying medicine. I enrolled at the University of Fribourg in Switzerland; I spent my first two years as a pre-med student. I spent the next four years in medical school at Bern, graduating in 1979.

Occasionally, when I read about cancer in medical textbooks, I thought of Ursula. I think now of how far we have come in taming the Hodgkin's disease that had taken Ursula from us. Back then, when Ursula was dying, very, very few afflicted with her disease could be saved. Today, most Hodgkin's victims can be spared through modern therapy.

Today, those afflicted with her type of cancer represent only a small number of all cancer victims worldwide. Millions of cancer victims remain without hope.

I tell this story on these pages therefore for all of them.

Meanwhile, I had decided to become a doctor. It was in 1972 that I met my future wife, Anne-Laurence.

From 1980 to 1982 I worked at first in pathology at the University of Bern. I performed autopsies and did biopsy analysis. I then went into internal medicine at the University Hospital of Bern and from 1982 to 1984 at the Waid Hospital in Zurich. After my return to the University of Bern hospital as a chief resident, I remained there for four years. Starting in 1985, I became intrigued with business and suddenly my career was at a crossroads.

2

On the Shoulders of Giants

I had always felt that business and politics were two of the key elements of any society and I felt I would like to know and understand more about both of them. Politics seemed an unrealistic pursuit for me; but I began to realize that business intrigued me more and more. My wife Anne-Laurence has an uncle named Marc Moret who in 1987 was the head of the large pharmaceutical company, Sandoz, headquartered in Basel. I found his industry fascinating—or at least I imagined it to be fascinating. I made a point of talking to him about my wish to drop medicine for business. He openly discouraged me. It was as if someone had thrown cold water on my face. "You are much better as a physician," he said, but if he were trying to discourage me, I was still intrigued about getting into the business world. He noted how many annoyances existed in business. I decided that he was simply frustrated because of negative publicity that Sandoz received after a large fire in 1986.

Then, in December 1987, a new head of Sandoz pharmaceuticals called Max Link, responded to my request for a meeting; I had wanted his counsel about entering an MBA program. He suggested that I get some on-the-job training in the Sandoz marketing and sales department in its United States affiliate. Thanks to his encouragement and optimism and the support of my spouse, I decided to make the change.

Two weeks later, I resigned from the hospital in Bern. I took up my work at Sandoz the following March and moved with my family to its American headquarters in East Hanover, New Jersey.

Those early days were a disaster. I had been assigned first to do marketing research and then to sales, but I really had no clue what the job entailed. I had gone from being a respected physician to someone behind whose back colleagues were saying, "What is that guy doing here? He doesn't understand anything. Why did he leave his medical profession?"

Slowly but surely I grew more competent and, learning new things every day, I began to enjoy my new job. I had a great experience in 1989 when I spent three months in a management development program at the Harvard Business School. It not only provided some fundamental business tools, but it also bolstered my self-confidence: brushing shoulders with veteran executives at HBS and finding that I could hold my own, I now felt that I could trust my judgment in business too.

Returning to Sandoz, I was given responsibility for the drug Sandostatin, which improves the symptoms caused by carcinoid tumors and treats acromegaly. Expectations for the drug were quite low but I was determined to make an impact. I loved the chance to manage a cross-functional team. When the product sold well, Sandoz gave me several products in the high-tech area. I remained in the United States for the next three years and thoroughly enjoyed my job, the country, and its people.

≡

It was around this time that scientists came up with critical breakthroughs in the understanding of CML, breakthroughs that would eventually lead to the development of Gleevec.

By the late 1980s, two researchers had identified the principal mechanism of CML. The two researchers were Dr. David Baltimore, then at the Whitehead Institute for Biomedical Research in Cambridge, Massachusetts; and today President and Professor of Biology at the California Institute of Technology; and Dr. Owen N. Witte, Professor in Developmental Immunology at UCLA.

In 1986 and 1987, Baltimore's research team published two articles in *Science Magazine* that pegged the Bcr-Abl protein as a tyrosine kinase, a type of enzyme that plays a critical role in regulating cell growth and division. In effect, Drs. Baltimore and Witte had now identified Bcr-Abl's cancer-causing properties.

It had not been known that the Philadelphia chromosome resulted in the activation of a tyrosine kinase. Now it had become apparent that a certain tyrosine kinase caused CML.

We cannot stress enough how important it was that scientists had identified what caused CML. In no other cancer had the molecular mechanism been identified in such detail. Knowing the cause meant that other scientists could now work on a drug that could counteract the disease.

Suddenly, the race to find the drug had a new focus—and a new opportunity.

$$\equiv$$

Just to digress for a moment, it is worth noting that the true beginnings of the genetic study of cancer came in 1960 when researchers discovered the Philadelphia chromosome. This discovery marked the first time that a cancer-related genetic abnormality—to be more specific, a chromosomal defect—had been identified.

It was then that a pair of researchers—Dr. Peter Nowell, of the University of Pennsylvania School of Medicine, and Dr. David Hungerford, of the Institute for Cancer Research—noticed strange behavior in the chromosomes of blood cells from CML patients.

The newly identified chromosome certainly had strange properties: One copy of a chromosome, later termed chromosome 22, was shorter than normal; an entire chunk of DNA was missing. The technology employed by the scientists was too crude at the time to tell which chromosome it was for certain. But the two researchers called it the "Philadelphia chromosome."

The Philadelphia chromosome, the abnormal-appearing human chromosome 22, results from a mutation that involves the swapping of genetic material between chromosomes 9 and 22. The abnormality results in a gene that produces an abnormal protein called Bcr-Abl. But all of this was not known at that time.

Although no one knows what causes this DNA change, it is clear that the resulting abnormal protein disrupts the bone marrow's normally well-controlled production of white blood cells. This "deregulated" production of white blood cells leads to a massive increase in their concentration in the blood, an indication of CML.

For the first time a chromosomal defect had been linked to a cancer. The fate of the missing DNA remained unknown.

The Tail and the Head

It took another 13 years—until 1973—for Dr. Janet Rowley, of the University of Chicago, to notice that CML patients had an extra clump of DNA on chromosome 9. When she put chromosomes 9 and 22 together, she discovered that the missing piece of chromosome 22 had shifted to chromosome 9 and a missing section from chromosome 9 had shifted as well to chromosome 22.

Those shifts are known as "translocation" in genetics parlance, a process in which a bit of genetic material from one chromosome swaps places with a bit from another chromosome. In CML, the "tail" of a gene (called "Abl") from chromosome 9 is translocated onto the "head" of another gene (called "Bcr") on chromosome 22, creating the Bcr-Abl oncogene.

The alteration of the DNA in the 9 and 22 chromosomes—the Philadelphia chromosome—was found to exist in 95 percent of CML patients. With the breakthroughs of the late 1980s identifying the mechanism for CML, the formula for the creation of a drug to combat the disease appeared in hand. All one had to do was to come up with a compound that could inhibit the Bcr-Abl oncogene.

But that would prove difficult.

Given the fact that no one thought it possible to develop a drug compound that could inhibit a specific tyrosine kinase, it appeared virtually impossible that a new cancer therapy could be found to combat CML.

What was needed was someone who believed in tyrosine kinases. But such people were few and far between.

The Bulldozer

His colleagues think of him as a real intellectual bulldozer and they mean that as a true compliment; but when you sit down with Alex Matter, he's a good deal gentler—and less noisy—than any bulldozer I know. But, make no mistake about it, once he joined the race that is the subject of our book, he never let up. He pushed and prodded.

A doctor and researcher himself, Alex had witnessed failure on the part of laboratory scientists too often. He felt sorry for these people, toiling in laboratories all their careers, and in the end, having nothing to show for it. He promised himself that he would try not to let this happen to him. From the very start of his career, Alex wanted to make an impact on the world. He wanted to do good deeds. Two of the greatest scientists, Louis Pasteur and Marie Curie, inspired him. As a boy of 12, he had dreamed that one day he would help in the discovery of some new medicine.

In the 1970s, Alex walked the corridors of a few hospitals in Western Europe, a resident physician trying to find his niche. He noted that most of the time the hospital doctors could save their patients' lives. Only one group, the oncology unit, seemed to have a hard time. Indeed, the physicians in oncology appeared helpless. Oncology was a field where simply nothing worked. People died.

Existing cancer treatments did little to prolong the lives of patients. All they seemed to do was make the patients' lives miserable; they certainly did not add to the quality of what remained of their lives. With its carpetbombing approach, chemotherapy killed everything in its path, both the good cells and the bad (the cancerous ones), turning patients into weak people with no hope for the future.

Alex felt strongly that someone needed to come up with a more sophisticated treatment. As I had with my sister, Alex had witnessed cancer take its toll on relatives and friends, some at an unreasonably early age. So for reasons that were both personal and professional, Alex Matter saw cancer therapy as an opportunity.

Certainly research into new cancer therapies was a long shot at best, but it offered a medical researcher in search of a cause the kind of challenge that seemed worthwhile. When Alex Matter voiced his thoughts to supervisors, they agreed with him wholeheartedly that cancer therapy, still in its infancy, was barely making a dent.

Alex, a native of Basel, Switzerland, had found his lost cause. He had spent years of study to arrive at this point. He had received medical degrees from the Universities of Basel and Geneva, and completed a doctoral thesis at the Institute of Pathology at the University of Basel. He then engaged in pathology and immunology research in Europe and the United States. He always believed that there was a great deal that one could learn from nature; here was nature coming up with a trick to selectively kill tumor cells. Might it not be possible to learn from nature, he asked himself, to kill cells? But all of his research efforts in that direction led to an impasse.

In 1980, setting up his own lab in France, he tried to do work on interferon. He shared the hope and excitement of other scientists that finally a drug had come along that might impact positively on cancer patients.

At the time, interferon was touted as The Solution to the cancer problem. But very quickly, it became clear that interferon had a positive effect on only a small percentage of tumors. Looking back at all the hype and heady optimism that surrounded the new drug, Alex felt annoyed and saddened at how naïve the scientific community had been toward interferon.

Interferon, to be sure, had its place in medicine: it certainly provided a better understanding of how biological agents could kill certain tumor cells; and it proved an important tool in combating viruses—but not tumors.

To Alex Matter's great disappointment, too few people were getting excited about cancer research. Ciba-Geigy, the large pharmaceutical company in Alex's native Basel, had shut its cancer research unit in 1980; management simply decided that the investment was not worth the paltry returns.

Then in 1983, one of Alex Matter's mentors, a man named Peter Dukor, let it be known that he would personally re-establish cancer research at Ciba-Geigy. He lured Alex back to Basel to head the old-new unit.

But the cancer research program was marginalized, left to its own devices by the senior management who dispensed the company's resources far more liberally to other research fields.

After a while, Alex discovered that it was marvelous not being bothered by his supervisors. He and the others in the unit could do more or less whatever they pleased. And what Alex wished to do was to exploit the revolution occurring in cancer research, a revolution prompted at first by the Watson-Crick DNA discovery, and later by subsequent progress in molecular biology research. It was, Alex recalled, a period in his life that was fantastic, so invigorating and so refreshing.

Suddenly, Alex sensed that he might be able to deliver on his dream of coming up with something spectacular in the field of cancer research. He genuinely believed that, based on the work of incredibly talented scientists in molecular biology, he could stand on their shoulders, and seek a specific compound that could keep cancer genes in check.

Thinking About Kinases

Alex was slowly turning his thoughts toward kinases. He had fallen under the spell of some Chinese physicians as well as some old medical friends from Basel, who were fascinated with kinases. The Chinese and the Basel doctors had stopped short of suggesting that certain kinases might be involved in the creation of cancer genes. They simply made the point that kinases had something to do with cell proliferation and had not explored whether it was worth trying to inhibit kinases.

Alex found their research and their perspective intriguing, for he knew that unregulated cell proliferation was a basic indicator of cancer. He had also observed that tyrosine kinases played some kind of role not only in normal cells, but also in cancer cells.

Not many pharmacologists shared his interest in kinases. To them, kinases were simply not worthwhile targets for compounds. The pharmacologists adopted the traditional view that any compound would have to aim at extracellular targets: the receptors at the surface of cells or things floating in the bloodstream. By honing in on those targets, there was at least a reasonable chance for the compound to arrive and interact.

To most other scientists, developing a compound that would reach the interior part of the cell where kinases were located seemed unattainable. The compound would have to cross a cell membrane to reach the kinase, considered an impossible hurdle at the time.

Alex Matter, however, still wanted to try.

He had done enough reading to make him believe that kinases might be his lucky charm. Once, while perusing a Japanese abstract, Alex came across a startling piece of information. Researchers had uncovered a certain natural compound called staurosporine, found in mushrooms, that was potent in inhibiting a wide variety of kinases.

This was indeed a breakthrough for Alex Matter. He delighted in the fact that nature seemed to be pointing the way to the solution of a medical problem: Nature had found a trick to inhibit kinases very effectively with this natural compound.

His colleagues could say what they wanted. For him, the race should focus on kinases.

Fascinated with the way Alex's mind worked, I once asked him whether fellow scientists had thought him crazy to pursue kinases. He looked at me as if to say "Am I responsible for the narrow-mindedness of my colleagues?" No one accused him of being crazy, he suggested, because no one even thought about targeting kinases with the idea of inhibiting them. It was just too outlandish.

Some of his colleagues were rather unpleasant. A few suggested that if he had been working under their supervision, they would not have allowed him near kinases.

≡

The year was 1985. Alex had now been in Basel for two years.

He needed someone who could assume the leadership in his lab in the study of kinases.

One person came to mind: Nick Lydon.

Alex was hardly optimistic about luring Nick to Basel. They had worked together in Paris and Lyons, and Alex thought very highly of the young researcher. But he had no idea whether Nick would be willing to take a very high risk and come work with Alex in Basel. When Nick replied positively, Alex thought the man had a lot of guts.

You Guys with Your Crazy Ideas

With Alex Matter and Nick Lydon going into the kinase business, the road kept leading from Basel to Boston and back again.

For Alex and Nick had something that the scientists in Boston wanted; and the Boston scientists had something that Alex and Nick needed.

It had the makings of a happy marriage.

Even before the scientists in Basel and Boston began to collaborate, began to move in the same direction, they had a common mission.

These were people who were not satisfied with just showing how things in the lab worked.

Each one of them genuinely wanted to do more than that. They wanted to help others. They wanted to do something in the lab that would benefit people.

Each fulfilled a vital role in the race. As they began talking to one another, they still seemed very far from a discovery.

There was Alex Matter, the man whose colleagues thought he was a bit crazy for spending so much time studying tyrosine kinases. But Alex, the bulldozer, would not be stopped.

More of a leader than a technician, Alex had turned to Nick Lydon to head up Ciba-Geigy's laboratory effort to study tyrosine kinases. Yet, even as Nick focused on these under-studied parts of the cell, Alex's young protégé knew that essentially he was doing basic research. To really help people, he needed someone to point the way, to suggest the diseases that Nick should target in his lab efforts.

That was the crucial contribution that Brian Druker, sitting in Boston at the time, provided in the nascent collaboration. It was Lydon who, at a crucial point in the project, would send a number of Bcr-Abl kinase inhibitors, including STI571, to Brian Druker so that he could test them in his lab.

To actually make a compound that could be effective in cancer therapy, there had to be a chemist, someone who could take the ideas of Alex Matter, of Nick Lydon, of Brian Druker, and devise an actual molecule that would keep a disease-causing tyrosine kinase in harness. Jürg Zimmermann was that chemist.

But Jürg needed some biologists to tell him whether he had accomplished the feat; scientists who could perform tests on his compounds and decide whether any of them might actually do the trick.

Enter Elisabeth Buchdunger.

The Team Forms

These then were the five people on our team as we began the race: Alex Matter, Nick Lydon, Brian Druker, Jürg Zimmermann, and Elisabeth Buchdunger. They were by no means the only ones who made contributions, but we will focus on their work.

They were indispensable to one another. Each took a part of the puzzle and placed it on the board in the exact right place.

But as they worked, they had no way of knowing if they would succeed. Indeed, had they been asked at any juncture during the struggle, they would probably have bet on their own failure.

Yet they carried on.

I have only admiration for their imagination, dedication and perseverance.

≋

As the head of Alex Matter's tyrosine kinase inhibitor program, Nick Lydon had the task of getting the right tyrosine kinases inhibitors into patient trials.

That led Nick Lydon to Brian Druker, who would play an important role in our story. He first came to the attention of our team in 1988.

It was three years into Nick's work at Ciba-Geigy. By this time, Alex Matter was strengthening his ties to the Dana-Farber Cancer Institute in Boston due to its expertise in tyrosine kinases. Thus it was that Nick Lydon traveled to Dana-Farber and came upon a post-doctorate medical oncologist named Brian Druker.

Druker wanted to return to the lab to start research into abnormalities that drive the growth of cancer cells.

He started studying tyrosine kinases in the laboratory of Dr. Tom Roberts at the Dana-Farber Cancer Institute. Druker knew that the most promising cancer for research was CML. Thus far, it was the only cancer—of more than 100 kinds—where the genetic cause was known.

Still, Druker remained skeptical when it came to focusing his research on tyrosine kinases. He simply did not believe that it was possible to create a compound that could target a tyrosine kinase selectively, i.e., that would inhibit a cancer-causing tyrosine kinase without proving toxic against any of the other 150 or so tyrosine kinases that are critical to the body's metabolism and cellular growth.

This was all just another way of saying that the race to find a compound that could impact positively on cancer was sputtering. Growing increasingly impatient, Druker would sit for hours with patients, knowing that there was little he could do for them. Soon after one died, he would write a note to the family promising: "I will remember what we couldn't do for your [mother, your daughter, your son, your husband]. They will motivate me as I enter my lab career, so maybe someday we can have more to offer."

Maybe It Can Be Done!

Late in 1988 something happened that would change Brian Druker's thinking.

It was then that a group from Israel headed by Professor Alexander Levitzki published an article in *Science* magazine suggesting that they had been able to selectively inhibit an Epidermal Growth Factor (EGF) receptor. Growth factors are chemicals that play a number of roles in the stimulation of new cell growth and cell maintenance. They bind to the cell surface on receptors. Specific growth factors, such as the EGF, can cause new cell proliferation.

Levitzki's article cast kinase inhibitors in a whole new light for Brian Druker. If it was possible to get some specificity (as Levitzki's group had done with the EGF), that meant one could now think seriously about creating an effective compound that would inhibit a specific tyrosine kinase.

Sometime in 1988, Druker suggested to Lydon that CML would be the ideal target for Lydon's research; Druker also predicted that

CML would be the first disease on which a tyrosine kinase inhibitor would work.

Until their conversation, Nick Lydon had made little effort to focus on the Bcr-Abl oncogene; it was not even on his list of compound candidates because inhibiting the Bcr-Abl oncogene, even if it could be done, meant creating a compound against a disease with very few victims.

Convinced that the Bcr-Abl oncogene was simply not that interesting, Lydon and the other Ciba-Geigy scientists had targeted other diseases for testing whether inhibitors could be created for tyrosine kinases.

But when Brian Druker suggested that a tyrosine kinase inhibitor that stopped the Bcr-Abl oncogene in its tracks would become a proof of concept for the whole molecular biology approach in cancer research, Lydon became interested. He made sure to add Bcr-Abl to his list of oncogenes against which he would test tyrosine kinase inhibitors.

As a new decade began in 1990, none of this appeared to make much difference toward the outcome of the race. No one really seemed much closer to developing a compound that would stave off the horrible effects of cancer.

Alex Matter and Nick Lydon continued to make preparations for the testing of various compounds, but they had not narrowed the list down to any manageable number yet. They simply had no idea what might work and what might not.

Meanwhile, Brian Druker had been forced to sever his relationship with the Ciba-Geigy scientists: Druker's Dana-Farber had just signed an exclusive agreement with Ciba-Geigy's rival Sandoz to work on signal transduction compounds. So from 1990 to 1993, Brian Druker lost contact with Alex Matter and Nick Lydon.

Druker, however, was making decisions that he hoped would bring him closer to solving some of the riddles of cancer research. He decided to transition his work in tyrosine kinase inhibitors from a rodent to a human model. Studying rodents would allow him to go only so far. The only way to truly prove that it was possible to inhibit a tyrosine kinase as a way of treating cancer was to engage in patient trials.

Meanwhile, in Basel some 100 researchers were assigned to the Matter-led labs at Ciba-Geigy, a far cry from the half dozen who had joined him in 1983.

Matter and Lydon had hundreds of compounds that they wanted to study; they of course would have liked to find a compound to combat CML, but more important, because of the larger medical need, they hoped to find a compound or compounds that would work on solid tumors of lungs, breasts, prostates, etc.

Their problem at the moment was that all of the compounds they had under study were weak inhibitors of tyrosine kinases, lacking potency, selectivity, and—most important—what pharmacologists call drug-likeness (the compound is properly absorbed orally; is non-toxic and undergoes no chemical transformation in the body; is stable in the stomach, etc.).

In the quest for the inhibitor, Matter and Lydon narrowed the list of hundreds of compounds to manageable proportions. To help them they hired a chemist and biologist.

Let's turn first to the chemist.

In 1990, Jürg Zimmermann joined Ciba-Geigy as a medical chemist in Alex Matter's Oncology unit.

I've only gotten to know Jürg lately, but I've tried to picture him in his laboratory in the early 1990s, writing down formulas on a piece of paper, turning to his computer to get the math just right. Even meeting him briefly, I can tell how much he loves chemistry, especially medical chemistry. Jürg's optimism is infectious, and he can't seem to get enough of the test tubes and Petrie dishes. But, for all his optimism and love of the subject, his task back then was daunting. He would need all of that optimism to see the task through. When he joined the race, there were only a handful of believers.

He had been born in a small mountainous village called Adelboden in Switzerland. At age 14, Jürg decided that his goal in life should be discovering something in pharmaceuticals that would help others. He has trouble remembering why he thought that, but recalls only that it seemed to be the most important thing in life for him.

Accordingly, he focused on pharmaceutical chemistry as a vocation. For a young Swiss chemist, Basel was the place to be: It was in that city along the Rhine River that three of the largest pharmaceutical companies were headquartered: Ciba-Geigy, Sandoz, and Roche. Jürg Zimmermann went to work at Ciba-Geigy on August 2, 1990.

≡

Assigned to work in Nick Lydon's protein kinase program, Jürg knew how high-risk the project was. At lunch breaks, colleagues teased him ferociously: "You guys with your crazy ideas." With so many protein kinases in the body, most needed for survival, it appeared hopeless that someone could create a compound that inhibited the Bcr-Abl oncogene selectively.

Jürg worked in tandem with the biologists at Ciba-Geigy, creating compounds that might become candidates for patient trials. He turned over compound candidates to biologists who made the final determination.

The biologists tested the compounds to decide whether any of them inhibited the Bcr-Abl oncogene. The process took many months. Days at a time, Jürg sat with pen and paper, drawing upon his knowledge of various chemical combinations. He worried about the shape of the compound—because it had to fit precisely into a so-called "pocket" of the Bcr-Abl enzyme. That "pocket" is filled by another chemical called ATP (adenosine triphosphate). The trick is to block the "pocket" with an inhibiting compound so that the Bcr-Abl oncogene does not work anymore.

Zimmermann worried about how the compound would interact as it raced through the different organs. At a later stage, he fretted about selectivity, making sure that the compound inhibited *only* the targeted Bcr-Abl oncogene, and left other good kinases alone to carry out their important functions.

Each week Jürg came up with some ten compounds. He then

turned them over to the biologists who did their testing over the next week or two. For Jürg, the waiting was nerve-wracking. He tried to work on other compounds, but his mind drifted to the biologists' labs and their testing. He grew impatient, but there was nothing he could do but wait for the results.

At first, the biologists came back with some disheartening news. Jürg was creating compounds that were indeed inhibiting the Bcr-Abl onco-gene; but they were inhibiting many other kinases as well. Such com-pounds would be highly toxic. It was back to the drawing board for Jürg.

≋

Alex Matter was at the all or nothing stage: He believed that either one of the highly ranked compounds had to work or everything he had learned about cancer was wrong.

Especially encouraging was the fact that scientists had already identified the link between the Bcr-Abl oncogene and CML. In most other cases of cancer, the signaling pathway remained unknown. What was more, usually two or more genes collaborated to form a full-blown cancer; but with CML, only one gene caused the disease, making it easier to come up with a compound.

But being easier did not guarantee success.

Alex understood how important it was for the lab team to succeed: If in this far less intricate case success proved elusive, how could one expect to succeed against other, more complicated diseases?

Getting the compound to work meant, in this case, a proof of con-cept, and Alex grew very excited at the prospect. He knew that failure was always possible. Indeed failure was all too common an occurrence.

Because it shared certain structural properties he found in many drugs on the market, Zimmermann was attracted to a class of com-pounds named phenylamino pyrimidines. His gut feeling told him that he could make something out of that class.

One of his tricks was to try to decipher secrets in the chemical structures of drugs already on the market—to try to find what they had in common: "If you can feel—it is indeed a feeling—how a good chemical structure must look for it to qualify as a drug, then you are a good medical chemist.

"You can make millions and millions of compounds but only a very small number of compounds have the properties to become a drug. All the millions and millions of bad ones have one problem or another: They can be toxic, they can be excreted very quickly, they can get destroyed in the liver, or they are not stable in the stomach. They might not be taken up in the gastrointestinal tract or they are not stable when you store them as a pill. They can bind to proteins in the blood. There are many, many hurdles! If you think about it you almost don't want to start the race. But history shows that persistent work can be successful."

≋

When Jürg Zimmermann arrived at work on August 26, 1992, he thought he was quite close to synthesizing the compound.

A series of compounds had passed muster in the biologists' labs on all the requisite tests save one: They were active against the Bcr-Abl oncogene; they were selective. They had even shown activity *in vivo* (animal testing).

The one thing keeping the compounds from getting a 100 percent stamp of approval was the issue of solubility. When animals were given the compounds, rather than show up in the blood, the compounds were excreted. Developing a compound that was soluble without reducing the compound's ability to bind to the "pocket" and without making it less selective was tricky. Jürg was not particularly optimistic that morning. How many times had he thought he had something only to discover that in the end, he had nothing? He had trained himself to dampen his enthusiasm. On that day, he turned over the promising candidate compound to the biologists cautiously hoping for the best. The focus thus shifted to Building 125, a laboratory in one of the buildings at Ciba-Geigy. The woman running the laboratory was Dr. Elisabeth Buchdunger.

It was Elisabeth Buchdunger's task to take the compounds that Jürg Zimmermann had prepared—including the one that he had synthesized back in August 1992—and have her team of biologists test them. She and her team had to find out whether these compounds remained selective when she tested them within complete cells.

Ever since Zimmermann had synthesized the promising candidate compounds the previous August, Elisabeth and her team of biologists had run test after test on them.

Then one day in early 1993, looking at the effect of one promising compound on the cellular level, she sensed that something interesting was occurring in her laboratory. Too used to failure, she would not allow herself to think grand thoughts.

On this day, however, her technicians appeared just a little bit more frenetic. They seemed to look at their equipment longer than usual. Not that these technicians were garrulous, but today they observed an odd silence.

What was going on?

Elisabeth asked them questions. She got only partial answers. The technicians liked what they saw but they wanted more time. They were trained to be cautious. When the signs were good, it seemed wise to be cautious.

Huddling over microscopes and other specialized equipment, Elisabeth and her team began looking at the way the compounds function in a cell.

To appreciate what Elisabeth's team was studying, we need to point out that a tyrosine kinase is an enzyme that phosphorylates—that is, it transfers phosphate groups to specific amino acids on a protein, allowing the protein to interact with the next protein.

In a normal cell, phosphorylation occurs when a specific need arises, such as cell proliferation or cell migration.

Under abnormal conditions a chromosomal translocation occurs that disrupts one gene that codes for a tyrosine kinase (e.g., Bcr-Abl). Separated from its own "head" and now under the control of another gene's "head" (Bcr) in the hybrid oncogene (Bcr-Abl), the tyrosine kinase becomes totally deregulated and remains "switched on" all the time.

The "always switched on" tyrosine kinase keeps on phosphorylating the substrate proteins in the cascade of reactions that tell the cell to divide, to migrate or to remain alive, even when there is no need for these functions.

This uncontrollable phosphorylation is a recipe for biological disaster, and its most sinister representative is cancer.

For a compound to work against CML, it must first of all control phosphorylation.

The crucial test that Elisabeth Buchdunger and her scientists performed took only a few hours. They studied the amount of phosphorylation that occurred when the compound acted on the Bcr-Abl oncogene.

Knowing the amount of phosphorylation—whether it is high or low—tells the biologists if a compound is actually inhibiting the oncogene's harmful effects.

The team incubated the cell cultures with the drug. Then they compared cells treated with the drug and cells that had not been treated.

What Elisabeth Buchdunger and her colleagues wanted to obtain was a reduction in the degree of phosphorylation—that would mean the compound was inhibiting the Bcr-Abl tyrosine kinase from working.

A few hours passed.

Members of the team made an important discovery; they found that the compound they had tested caused a reduction of phosphorylation in the cell.

They reported their findings to Elisabeth Buchdunger; a few minutes later there were smiles on their faces.

There were still many tests to go—on cancerous cells extracted from patients, on animals, on human beings during clinical trials.

But a moment of great medical importance had just occurred.

A compound had been found that indeed inhibited the activity of the cancer-producing Bcr-Abl oncogene.

Elisabeth Buchdunger brought the happy news to Jürg Zimmermann. He was elated.

3

The Diamond

In the spring of 1993 the compound that Jürg Zimmermann had synthesized and that Elisabeth Buchdunger had tested attained "drug candidate status," i.e., the preclinical data looked promising.

No one opened a bottle of champagne. No one thought the race was over. It was far too early for celebration. The goal was to move ahead speedily toward patient trials.

Now came one of the biggest challenges: Finding a physician in a hospital willing to test the drug on patients. Over the next few months Nick Lydon turned to a few hematologists to ask if they would test the compound in patient trials. The answer always came back no. The hematologists had reasonable excuses: They were busy with other things. There were already drugs that could be used to prolong the lives of CML patients, however briefly. The truth was that agreeing to patient trials on STI571 offered little in return. CML simply had too small a patient population.

The scientists in Switzerland felt they were at a crossroads. Thus far they had found no one who was prepared to take the compound to the next step. At the same time, marketing managers were discouraging management from supporting the compound. It could never be a money producer; best to leave it alone.

One Last Hope

Nick Lydon did not want to give up. Nor did Alex Matter. They had one last hope, however small it seemed. Around that time, Brian Druker traveled to Basel for meetings with Nick Lydon and his team. He too was at something of a crossroads. He was making plans to move from Dana-Farber to the Oregon Health and Science University (OHSU) in Portland, Oregon. With the move he hoped to focus on finding scientists who had the best inhibitor for Bcr-Abl and to bring that inhibitor to patient trials.

Nick Lydon showed the data on STI571 to the visitor, suggesting that this could be the compound that might have a powerful effect on CML. Jürg Zimmermann, present at the meetings as well, assumed that Brian Druker would, as others had, turn down the offer to undertake further studies on the compound.

But Brian Druker had his own hopes and dreams, and fortuitously, a common ground with the scientists.

As he sat and listened to Nick Lydon talk about the wonders of the new compound, STI571, Brian Druker sensed that he and these scientists were the perfect match: the scientists wanted to make compounds to inhibit tyrosine kinases; and Brian Druker had the experience in developing reagents for detecting activity of tyrosine kinases.

To their surprise, Brian Druker reacted very positively: "We must do something with the compound," he told the group in an excited voice. Druker believed that STI571 was clearly the best at killing CML.

Jürg Zimmermann sat across the table from Brian Druker and was filled with joy at the validation that the visitor had given to his work in the laboratory.

By August 1993, the Swiss-based scientists delivered the four best compounds to Brian Druker who carried out protein, cell and animal studies on the compounds using his CML model systems. STI571 was one of the compounds tested.

The real breakthrough came when he found that the STI571 compound killed a CML cell without harming the normal cells.

Druker completed his tests the following December. He liked what he saw and he communicated his enthusiasm to Nick Lydon: "I think we've got something here. We're getting some great results. This looks interesting. This is as promising as it gets in anybody's career."

But it was still a long way to patient trials. That would be the ultimate test.

In February 1994, Brian Druker presented his data for the first time formally to the Swiss-based scientists.

Brian Druker's results showed that the compound inhibited 90 percent of the leukemia cells *in vitro*.

This was dramatic news.

It still did not offer proof that the compound would help humans afflicted with CML, but it was certainly a major step forward.

Brian Druker told those attending the meeting that he felt the compound was a "diamond."

On the basis of that presentation, decisions were taken to proceed with further tests, all leading to patient trials.

The testing process of the compound, vital for progress, tested the nerves of the scientists as well. For they knew that one false step, one bad result, one instance of excessive toxicity, and the project could die on the spot.

Right now, the goal for the scientists was to get to patient trials.

That made the next step, the animal studies, set to begin in 1995, crucial.

If the animals reacted poorly to the compound, the scientists might have to declare the compound a failure.

But if the animals made their way through the tests, the scientists could learn a good deal about the proper dosages to give human beings.

All of the animal studies went reasonably well.

Still, no one was pushing the animal studies very aggressively. There was little enthusiasm from the senior management at Ciba-Geigy. No one would have been surprised to learn on any given day that the project had been scrubbed and resources put into other projects.

≋

While the chemists and biologists were testing the compound in their labs, I was working for a rival pharmaceutical firm. In June 1992, Rolf Soiron, the new Chief Operating Officer of the pharmaceutical business of Sandoz in Basel, Switzerland, asked me to move from the United States to become his personal assistant. He needed someone who knew the American market.

I became the leader of a six-month project on how to reorganize our product development program. Working also on the project were Jörg Reinhardt, our current head of development; Jürg Meier, then the worldwide head of research of Sandoz Pharma; and Henri Vanni and Thomas Wellauer of McKinsey & Company. Then I was promoted to global head of product management, because, even though I was responsible for sales, I had no formal authority over the managers in the various countries.

Later when Rolf Soiron asked me to become head of development, I was reluctant—worried that I would have trouble with a job that seemed to have less to do with business and more with research. But I loved the job. After a brief period as COO when Rolf Soiron left the company, I became the CEO of Sandoz's pharmaceutical business in 1995. When the merger occurred between Sandoz and Ciba-Geigy in March 1996, I was tapped to be the CEO of the newly merged company. I became Chairman as well in 1999.

The merger had been arranged when Marc Moret, then chairman of Sandoz, invited Louis von Planta, the ex-chairman and CEO of Ciba-Geigy, for a lunch to discuss the possibility of a merger between the two companies. Highly respected as the "grand old man" of Swiss industry, von Planta facilitated the first meeting between Marc Moret and Alex Krauer, chairman and CEO of Ciba. Moret wanted to merge the two companies as a way of repositioning the new company as a stronger industry player. The new company became the second largest drug maker in the world, holding a 4.4 percent share of the world pharmaceutical market, just behind then GlaxoWellcome's 4.7 percent; it was twice as large as its nearest rival in agrochemicals. The new company had a market cap of $75 billion.

The motto of the new company was "new skills in the science of life." Our goal was to make Novartis the world's leader in selected therapeutic areas.

Effects of the Merger

One of my tasks, after taking over my new function in Novartis, was to meet with the new research teams, including Alex Matter, who, as a result of the merger, was given broader responsibilities, taking charge of a much larger oncology unit. Before the merger, there had been a Sandoz oncology unit and a Ciba oncology unit. After the merger, the oncology unit under Alex Matter was twice the size; he had muscle that allowed him to do things that he could never have dreamed of before. Now he and his team had an international presence and they inherited from Sandoz a significant collaboration with Dana-Farber.

I had not known Alex at all before the merger and when I first met him in these post-merger meetings, he demonstrated his candor, complaining about the merger-related changes, the bureaucracy and inefficiency. He was quite difficult to deal with. He had set the highest standards for our oncology research and he had personally led the oncology unit down a path of innovative pursuits. At those early meetings, he talked a bit about a new compound that his people were studying that appeared to have some extremely interesting properties.

At first, I listened without much of a response but I became more and more impressed with this remarkable scientist. And, if he got behind a project enthusiastically, I found that one could not ignore him; it was that way with STI571. I believed in research and development as the crucial backbone of a pharmaceutical company. But I could not possibly know at such an early stage in the development of a drug what was going to work, and what was not.

I also know all too well—though trained as a physician—I cannot add scientific value to the work some of these people do. They are more sophisticated; they know much more. They keep up in science in a way that goes far beyond what I could ever do. So the questions I have to ask are:

Do the things that they say sound right and logical?

Is their thinking coherent?

Do they have the motivation and enthusiasm and persistence which is needed to conduct good research?

Are they open toward the outside? Can they absorb outside ideas or are they wrapped up entirely in their own world?

So I had to make judgments about the scientists. Who seemed to have more credibility? Who seemed less likely to overhype a compound?

The more I got to know Alex Matter, the more I trusted his competence and judgment.

And so I made clear to Alex that if he felt that strongly in favor of pursuing studies on this compound, he had my support. It was not a matter of my turning over more financial resources to him. It was more that I was saying to him, "Go for it," and that apparently meant a great deal.

Alex will tell people that it was my personal interest and enthusiasm for the compound in the post-merger days that got the project really moving.

He liked to say that when he promoted the compound before the merger, people would say dismissively, "Oh, it's him again. He's at it again." They did not take him too seriously. Oncology was not taken too seriously. Why should it have been as its scientists constantly boasted of finding the next wonder drug, only to watch it fizzle?

But if one listened to him, one could sense the logic of the approach and I knew that he was a realist, not a dreamer. Working with Alex Matter turned out to be a great experience.

≋

In April 1996, Brian Druker began planning for patient trials at the intravenous stage for the end of the year, provided the animal studies continued to do well.

Finding patients for Phase I patient studies tends to be a fairly large problem, and Brian Druker wanted to get at that problem right away. The publicity that the compound was attracting helped in finding patients.

On May 1, 1996, Brian Druker published what was then the first paper on STI571 in the *Journal of Nature Medicine.* There was some newspaper publicity surrounding the article. *The Oregonian* published a feature story and the Associated Press sent out a national story. Patients began calling in the wake of that mild publicity. Druker started to compile a list of patients for trials.

But then the dogs in the studies began developing blood clots at the point where the catheter tips were placed. It might have been caused by some mechanical problem with the catheters or it could have been caused by the compound itself. Whatever the case, there would be no patient trials until this matter could be resolved.

A heavy gloom cast itself over the scientists.

The options were not pleasant: either repeat the studies or wait for the toxicology studies on the oral formulation of the compound, and those studies were six months behind.

≋

Even with the degree of support that I was showing the project, there were enormous pressures on Alex Matter's team, pressures that spelled trouble for the compound seemingly at every turn.

First, Alex's oncology unit was still relatively unsuccessful and accordingly it still had little credibility among his supervisors in the corporation.

Second, in November 1996, the compound began to show high toxicity in the animal trials. It was by far the worst setback of the entire project. Studies on dogs given the compound intravenously were showing that the animals had suffered severe toxicity. The animal trials had to be halted there and then.

A genuine fear arose among the scientists that the drug was simply not worthwhile.

Already, there was plenty of talk from the marketing people that CML simply had too small a patient population and so any compound to combat it would never make any money; it was just a waste of time for the company.

But I suggested that if a compound proved promising, if it seemed likely to be medically significant, it made no sense to halt the research because of weak commercial projections. We can always find a commercial solution, I said at the time. We cannot let ourselves be imprisoned by myopic marketing considerations.

There were indeed many pressures. There were many chances for failure. "One has to realize," Alex Matter liked to tell me, "that in every single company, the default setting is that compounds like this one falter; they will not survive."

The one argument that Alex always used about STI571 was that it offered a proof of concept. For him, this took on the tone of a crusade. He could not believe that the concept would not work. It simply had to.

He wanted desperately to get to clinical trials; he worried constantly that the day would not come. It had happened so often in the past: People would find reasons to kill something before going to patient trials. There always seemed to be a dozen good reasons.

But Alex Matter believed in this compound, believed that at the very least it would provide a proof of concept; and if all went well, it might turn into an actual cancer-fighting drug. He knew that people were tired of hearing from him. But he did not care. He was convinced that he had right on his side. The others would simply have to get used to him. Once Alex told me his strategy: "Clearly it's fighting every day. You never let go. It's like a mad dog. You take that bone and you do not let go. You have several projects. This was not the only one. I always tried to have a whole panoply of things that were cooking and were at different stages of development; to have a portfolio of programs and of production; not just to bank on one." But this compound was a pet project and he could not let go of it.

Nonetheless, with the negative results of the animal studies in late 1996, the project went into limbo.

Brian Druker found the delay unfathomable. He was no longer sure the project would make it out of the Novartis research labs. He was eager for patient trials. And nothing dissuaded him from the belief that the compound deserved a patient trial, even with the problems with liver toxicity, which he thought of as a nonissue.

He knew that, if a patient ever developed a liver problem during patient trials, he would simply stop the drug. That was the point of patient trials—to find out whether such problems would arise. How could anyone even know there would be any toxicity in human beings—at least until they were tested?

Maybe he would look a bit cavalier, but he simply did not believe that the liver toxicity issue should have to kill a promising and potentially life-saving cancer drug. Yes, the problem had to be watched carefully and it could not be taken lightly, but it should not kill the program.

And yet that was exactly what was happening; if not killing, at least delaying things.

Lurching Forward

Throughout the early part of 1997, the project remained in limbo due to concerns about the animal toxicology findings. The scientists felt little urgency to resolve the toxicity issues.

But somehow, there was just enough motivation to keep the compound from slipping through the cracks. Although repeat studies still showed problems, after much discussion the potential risk-benefit ratio favored starting the patient trials.

But always there were speed bumps.

One had to do with how to give the eventual drug to a patient. At this time, the only way the compound could be administered was intravenously. To Brian Druker, this approach made little sense: If you are going to inhibit a tyrosine kinase, you will have to inhibit it on a continuing basis, he told Nick Lydon and the others. A patient would have to either visit a hospital regularly or be admitted to a hospital constantly if the drug were given intravenously. Nick Lydon and Peter Graf, another Novartis scientist, worked hard, accordingly, to turn the compound into an oral formulation.

Peter Graf was working in Switzerland in 1984 as an analytical chemist and pharmacokineticist. He developed an analytical assay method and protocols for the evaluation of the basic pharmacokinetics in animals—which ultimately resulted in the oral formulation of STI571.

There were rumors that the pill would not work orally because it would not dissolve. But the scientists plunged ahead, performing a test on rats and dogs that indicated that the compound in the form of a pill could be absorbed.

Other speed bumps arose. One scientist complained that the compound tasted bitter. The presumably bitter taste was reason enough to some scientists to avoid oral development of the compound. To decide once and for all, Peter Graf organized a tasting session in his office with three other scientists on the project team. The verdict came in: thumbs up. The taste was acceptable or, as one Novartis scientist suggested, "nothing which could not be masked by orange juice." (Indeed, patients have found the drug extremely bitter but even children take it easily when diluted with apple juice.)

Another speed bump had to do with the complaint that the compound was very corrosive and might destroy the machines that would manufacture it. The project scientists believed that this was yet another effort at stymieing oral development of the compound. When the decision was made to develop capsules, the problem simply disappeared.

More toxicity studies were ordered up on monkeys in the belief that it made sense to pay close attention to toxicity results in a species that most closely resembled humans.

The monkey results showed no liver toxicity at reasonable doses of the compound. At high doses, it showed toxicity. Nonetheless, the scientists thought it acceptable to move on to human trials.

Despite finding toxicity in the rats and dogs, the scientists decided not to change the chemical make-up of STI571. It had already been optimized and whatever changes they made would require starting all over because the drug was so complex.

≋

It had been a long road, decades in fact. But now the time was approaching when the drug could be tested on human beings. The scientists remained cautious. They knew that the drug could show promise in animals, but easily fail in humans. Many drugs did. But there was nothing more they could do. All they could do at this stage was to hope.

Patients and Doctors— The Awakening

The Moment of Truth

EARLY 1998

In the testing of a drug, the very first human trial is a critical step. It is a moment of truth.

Only when patients actually take the drug can anyone know whether they have a substance that is tolerated and effective.

Human trials have to be very exacting. Only certain patients can be permitted into the trials. Because the drug being administered is by definition new and experimental, the patients have to be protected against adverse effects. Dosages of the drug have to be administered with extreme care. Every aspect of the trials must be monitored and proper records kept.

Don't Harm the Patient!

It should be a time of high drama. But ordinarily it is not.

These then are the Phase I patient trials. They normally do little other than test dosages and metabolism on patients. To be rather brutal about it, the main object of these early trials is to make sure the drug can be given in specific doses to test efficacy and safety in larger patient trials.

The drama is usually left for later trials. No one expects patients to get out of their hospital beds and return home in a few days after taking the drug. Drugs do not usually work that quickly on patients.

So it was with the human trials for STI571—not much excitement in the beginning, not much drama. But the doctors were very curious. The scientists had a good feeling about this drug. It had done all the right things in previous laboratory tests. It represented a whole new approach to cancer therapy.

The doctors were holding their breath and hoping.

It was early in 1998 that we at Novartis gave the green light to move ahead with the patient trials on STI571.

Brian Druker and his colleagues were actually meeting and discussing the design of these trials as early as February 1994; back then Druker presented Novartis with a list of issues to be resolved before beginning patient trials.

First of all: Who should get the drug?

Should it go to newly diagnosed patients or to those diagnosed some time in the past?

Standard procedure in cancer patient trials calls for first giving the drug to patients with the worst prognosis. It is considered unethical to give the drug to newly diagnosed patients for they may have a decent prospect of responding to treatment using standard therapies. And the trial drug is, after all, experimental—and not yet tested on human beings.

As we noted earlier, with CML patients, the average survival rate with chemotherapy treatment had been four years. The average survival rate with interferon improved to six years. It would have been unethical to give the compound to patients who were newly diagnosed with CML, because if it did not work it might deprive these patients of an additional four to six years of life. Moreover, Phase I cancer trials routinely include patients who had failed first-line therapy. For CML patients, this essentially means patients who have failed therapy with interferon.

Fatalities at the start of patient trials are rare because such low doses of the drug are given; but no ethical committee would have tolerated newly diagnosed CML patients in the Phase I trial with STI571.

Another group of CML patients, those who have arrived at the blast crisis (third and final) phase—when the disease transforms into an acute leukemia—were also not good candidates for Phase I patient trials for a different reason.

It is generally believed that blast crisis patients have mechanisms other than the Bcr-Abl tyrosine kinase driving their leukemia; thus negative results yielded from their taking the drug might be a false negative and might not be predictive for an excellent response in patients at an earlier phase of the illness.

Ruling out newly diagnosed and blast crisis patients, the trial team decided to take only CML patients who had failed interferon: the outlook for these patients is grim since the majority enter blast crisis within one or two years.

≋

The trial team had to wrestle with a number of thorny issues before starting the patient trials.

Perhaps the thorniest was dosage.

The tendency is to start patients on low dosages of the drug to avoid toxicity, but not so low that it would take years to determine if the drug was working.

The investigators had strong concerns that too high a dosage might kill off the patients' bone marrow along with the cancer cells within the bone marrow. Perhaps, they thought, some bone marrow from each patient should be frozen, and then given back if needed. In the end, they decided not to freeze the bone marrow, figuring that if they began with low enough dosages, they could fix the situation as they went along.

Then there was the question of how often to give patients the drug.

The drug could be given once or twice a day, 365 days a year; or, as with other cancer therapies, it could be given in "pulses"—one week on; three weeks off; one week on.

Based on the animal studies, the investigators at first thought it might be necessary to give the capsule two or three times a day to patients. Then they learned from the data of the first six patients on the Phase I trials that the half-life of the drug was five times longer in humans than in animals. That discovery meant that the drug could be given once a day, not more often.

That was good news because it is much easier for patients to remember to take one pill a day rather than several.

≡

Brian Druker was designated lead investigator; it had been a long struggle for him to get patient trials. He considered getting Novartis to agree to a starting date his greatest challenge; the trials themselves, he let us know, seemed much easier.

Druker asked Charles Sawyers of UCLA to be the second of three investigators on the project. The two men had collaborated on some scientific work earlier. Druker chose Sawyers because both had a laboratory interest in CML and Bcr-Abl signaling and because Sawyers was, like Druker, a medical oncologist. Sawyers had been working on the effort for 11 years and could not quite believe that the time had come for human trials. Like Brian Druker, he felt nervous: "There's this strong sense that this is going to be very complicated, that we shouldn't get our hopes up. Human cancer is so different from the laboratory models in cancer. We think we understand the molecular biology of the disease, but nature always humbles you." We at Novartis were concerned that there would be enough patients and so we asked Druker to add a third investigator, Moshe Talpaz of the University of Texas MD Anderson Cancer Center in Houston. Anderson had one of the largest CML patient populations in the world. Druker was hesitant to include Houston at first, but eventually realized that Moshe Talpaz's clinical expertise would be useful. Working with Druker, Sawyers and Talpaz, on our side at Novartis was Dr. John Ford, who was the International Clinical Leader.

Magic Cancer Bullet

Target Date Set

The Phase I trial was to start in early summer.

Brian Druker, Charles Sawyers, Moshe Talpaz and John Ford met in the first week of April when they put the finishing touches on the design.

Druker went into the Phase I studies nervous, excited, reasonably confident about the results, but ultimately, uncertain about what would happen. He had been through this before in 1996, preparing for a Phase I trial for the compound, only to be disappointed at the eventual delay.

He had been working on this compound for five years and had pushed hard to get it to patient trials. He knew how much of an investment we at Novartis had put into the compound. If for some reason it did not work, he worried that we would not speak to him again—after "wasting" hundreds of millions of dollars.

He felt he should temper his excitement. He might have to go through disappointment again. All sorts of things could happen. That nagging problem of the liver toxicity could show up in patients. Or there could be some other new toxicity. The compound might not penetrate into the patients' cells. Or there could be new, unknown problems.

But he truly believed that he was on to something big. He could not predict what the side effects of the compound might be on patients; but he had a strong feeling that the concept was valid. He genuinely believed that the Bcr-Abl oncogene was the key target to be inhibited, and therefore as long as they could get to the right dosages, they could control the disease. Things seemed pretty simple to him: As long as you have the right target, if you shut down the kinase activity, it should work.

The investigators sought patients, either from their respective hospitals or through referrals.

The First Patients

By June 1998, Druker had the names of 12 patients on his own list. Using one a month would be enough to sustain him throughout the Phase I trial. His original thought was to have 50 patients in the study. Druker and Sawyers, wanted to make sure they had a reasonable number of patients from their hospitals; otherwise, Anderson, which had 300 CML patients a year, would have had a substantially larger share of the patients enrolled in Phase I.

It was agreed that each of the three centers was to get one of the three cohorts of the doses. (A cohort is a certain number of patients who receive an identical dose of the trial drug.)

The Phase I trial for STI571 eventually enrolled 149 patients—quite a large number for a Phase I study. Normally, Phase I cancer trials include no more than 30 to 50 patients.

Why was this study so large?

First, the findings in CML patients were so positive that the team decided to add some blast crisis patients to the study to see if STI571 worked on them as well.

Second, the investigators—to their surprise—did not run into dose-limiting toxicity, which meant that there could be a larger number of dosage cohorts. In the end they had 14 dose cohorts.

To be a part of the study, the patients had to have an elevated white cell count; otherwise there was no way to know whether a drop in their white cell count was due to the drug.

However, CML patients who had failed interferon treatment might nonetheless have normal white cell counts because of the hydroxyurea (a relatively mild type of chemotherapy) that they continued to take. To make sure they were eligible for the STI571 Phase I trial, patients were taken off of hydroxyurea and admitted to the trial once their white cell count had reached 20,000 per cubic millimeter.

Some patients saw their white cell counts rise to the required level within a day or two; others required four or five months. One patient spent six months living in a camper van in Los Angeles waiting for her white cell count to climb sufficiently. Eventually, the team allowed her into the study even though her counts were slightly below the desired level.

≋

Patients were chosen on a first come, first serve basis.
The three criteria for inclusion on the Phase I trial were:

1. The patient had to have CML in the chronic phase that was positive for the Philadelphia chromosome.

2. The patient had to have failed interferon therapy.

3. The patient had to have a white cell count of at least 20,000 per cubic millimeter.

In the end, there were 84 CML patients in the chronic phase and 59 blast crisis patients. Six children, all who had been resistant to interferon, were added to the study, some in the chronic stage, some in blast crisis.

JUNE 1998

An Update on Judy Orem

Back in May 1996, a friend of Judy Orem's was listening to the radio and heard something that she thought would interest Judy. She phoned her right away. On the radio, people were discussing leukemia and Brian Druker's name was mentioned. It seems he had just been given a translational grant from the Leukemia and Lymphoma Society to get a new research drug ready for human trials.

Human trials were a year or two away, but just being in the wings gave Judy hope.

The friend had even called up the hospital at the Oregon Health and Sciences University and obtained Druker's phone number, which she gave to Judy. The friend left Judy's name and phone number with Druker's office.

Within a day or so, Drucker and Judy were speaking on the phone.

She told Druker her condition, how she had been diagnosed in 1995, and that she was on interferon. In response, he said he was just gearing up for patient trials but it would take some time. He would keep her name on file and would stay in touch with her to let her know how he was progressing.

Judy made sure not to sound desperate on the phone. She was, after all, not in dire straits of needing anything; she was tolerating the treatment she was getting.

Brian Druker kept in touch with Judy Orem on a regular basis after hearing from her in the spring of 1996 that she wanted to get on the patient trials for STI571 as soon as possible.

When the Phase I study began in June 1998, Judy had just become eligible. She had just received evidence that, even after taking Ara-C for six months, her treatment was no longer working: Judy's bone marrow biopsy showed that she was back to 100 percent Philadelphia chromosome positive.

She was still in the chronic phase, but her doctors were now predicting that she might enter the accelerated phase at any time from the next six months to two years.

She looked again seriously at having a bone marrow transplant. But when she checked with doctors, they now said her chances of survival had dropped from 50 percent (before) to only 5–12 percent. She decided against the transplant.

Her physician could offer nothing further: "There's nothing else we can do," he told her. He thought keeping her on interferon made sense because at least it kept her white cell count under control.

Judy stated the obvious: "That's not a terribly exciting way to live."

JUNE 22, 1998

The Trials Begin

The study began on June 22, 1998 with the first patient, a minister from Bakersfield, California, taking the drug at a dose of 25 milligrams daily at Charles Sawyers's center in Los Angeles. Because he was the first patient in the trials, he was hospitalized and while he took the first pill, his family took pictures of him.

JUNE 25, 1998

A Second Patient

A second patient, Bud Romine of Telmec, Oregon, began taking the same dose on June 25. He was Brian Druker's first patient in the Phase I trials. Bud had contacted Druker after seeing *The Oregonian* article on the compound in April 1996.

Druker feared Bud Romine's disease would get worse. For one thing, Bud had wildly fluctuating blood counts: he would go from 150,000 to 10,000 in a few weeks. But he was in the chronic phase so Druker felt he was a good candidate for the trials. Bud certainly met the criteria: he had failed standard therapies. That meant that his blood counts had increased during interferon therapy; or he had no cytogenetic response after a full year of interferon.

JULY 16, 1998

How Could This Happen So Soon?

On day 24 of continuous daily treatment, something happened. The minister's white cell count suddenly normalized. The other two patients on the study had no response at all at 25 milligrams.

John Ford was vacationing at the time. He put in a routine phone call to Brian Druker and heard the news. Could this be? Ford asked excitedly. How could this be happening so soon?

Their excitement lasted no more than a week. After that, the minister's white cell count rose again.

It took the team a while to figure out what was happening.

STI571 had not been the cause of the patient's white cell reduction.

What Druker and Sawyers were interpreting at first as a positive reaction to STI571 was nothing other than the usual oscillation of the white cell count over a several month period, a phenomenon doctors call "cycling."

The phenomenon is a well-described occurrence in CML patients and in healthy people as well (though the counts oscillate within the normal range in healthy people).

The team finally concluded that all that they were seeing was this natural "cycling" of the disease. Naturally, they experienced a certain amount of disappointment.

The minister took the pills at 25 milligrams for six weeks but his white cell count continued to drift upwards. He was eager to stay on the trial and the rise in his white cell count was not life threatening so there was no reason to shift his therapy to chemotherapy. Eventually Sawyers told him that, with the drug not working, he would have to go off the trials which he did on day 147 when his white cell counts were three times higher than baseline levels. He went back on hydroxyurea.

The second and third patients in the OHSU trials came from California.

Typically in Phase I trials, physicians begin patients on low dosages of the drug and then increase the dosage steadily in new cohorts of patients (three at each dose level). Each patient remains at the same dose level that he/she was taking upon entering the study. Dose escalation for an individual patient is rarely allowed.

John Ford had proposed a dose escalation scheme that was quite aggressive: the first patient was to be given 25 milligrams a day, with seven cohorts at increasing dose levels up to 800 milligrams a day.

However, because of findings in the animal studies, the Food and Drug Administration insisted on a more cautious dose escalation scheme rising no further than 400 milligrams daily.

Every four to six weeks, John Ford and the three investigators held teleconferences to assess the progress of the trial. The investigators sent Ford the patients' white cell count levels and he would hang the results on his office wall. It was a slow process. Sometimes it seemed as if nothing was happening. For some parts of the trial, nothing was.

LATE SUMMER 1998

Three Patients at 50 Milligrams

Three patients took 50-milligram dosages in July and August. Their white cell counts rose too high, and it was deemed unsafe to continue them on the drug.

AUGUST–SEPTEMBER 1998

Four Patients at 85 Milligrams

Four more patients took 85-milligram dosages in August and September.

To recap the first few months of the Phase I trials: A handful of patients continued taking the drug during August and September 1998 at doses of 25, 50, 85 and 140 milligrams daily with little convincing change in their white cell counts other than the cycling phenomenon.

SEPTEMBER 1998

First Glint of Progress

The first sign of any progress came in September 1998.

It was then that those taking the 85-milligram dosage saw their white cells stabilize. The investigators felt a bit encouraged, but only a bit.

OCTOBER 1998

The Gynecologist Patient

In October the doctors raised the dosage to 140 milligrams and all three new patients at that dosage had their blood counts fall radically, one to normal.

One of those patients was Dr. Nora Flanzbaum Friedenbach, a gynecologist who was born July 3, 1956 in Buenos Aires. She was the third patient to enter the Phase I trial in Houston. She had gone to medical school in Argentina. She had a husband, who was an accountant, and three children: twin girls, both 19, and a boy of 12.

Nora was diagnosed with CML in 1991 when she was only 35 years old; she went on interferon but suffered terrible side effects, including fever and muscle pain, and was bedridden for a great deal of the time. She had also tried hydroxyurea but her white cell count was difficult to control despite high doses of this medication.

Considered resistant to interferon, she qualified for Phase I; as required, she stopped taking hydroxyurea, and the day after, her white cell count shot up to 200,000, suggesting that she was heading toward the accelerated phase of the disease.

At the 140-milligram dosage, she clearly responded to the drug though her greatly elevated white cell count did not normalize completely.

After just three months in Houston she was understandably impatient with the program. She missed her husband and children in Buenos Aires. She just wanted out of the routine. "I had an emotional problem. It was very terrible for me," she commented in August 2001.

In January 1999, she voluntarily took herself off the drug that was arresting her CML and returned home.

Dr. Friedenbach presented a serious problem for the investigators: until her departure, they had treated only a dozen or so patients in the Phase I trial and here was a patient who had shown a positive reaction to the drug picking up and leaving in the middle! It was impossible to manage her therapy so far from Houston. At the time, she represented one of the most serious disappointments of the Phase I trials.

After her return to Argentina, her disease grew bad again and in September 1999 she sought out Moshe Talpaz. By then he and the other investigators had given 300-milligram doses to several trial patients for many months and they knew that this level was both safe and effective.

Nora asked to return to the trial and the investigators agreed; she was given 300 milligrams a day of STI571. But this time she was allowed to spend most of her time in Argentina. John Ford noted: "We took a considerable risk in allowing her to continue therapy in Argentina, so far from the study site in Texas that was monitoring her therapy. We felt able to do this because she was medically quali-fied and we were confident that she would recognize any potential toxicities."

She has been in a complete hematological response ever since: Her cancer cells have totally disappeared from her blood.

In August 2001, when we spoke with her, she was taking between 400 and 800 milligrams a day of STI571. She was brutally honest about her current state of health: "I have at this moment a hemato-logical remission; I don't have a cytogenetic remission. But the side effects are very minor; I have a little edema around my eyes; a few muscle cramps; but it doesn't compare with the dreadful side effects of interferon." Other than that, "I feel well."

John Ford adds, "When Dr. Friedenbach first went on the study in 1998, she had early signs of disease acceleration, as shown by the fact that her blood count was difficult to control. Given this back-ground, it is unlikely that she would still be alive today had she received standard therapies."

≡

Judy Orem had her doubts about entering the Phase I trials, worried that the drug might prove toxic and kill her instantly. She might have preferred waiting for a Phase II study, but she really did not have that kind of time.

When she saw Brian Druker in October 1998, he asked her if she wanted to join his Phase I trials for STI571. He calmed whatever fears and anxieties she felt about joining an experimental drug program. We are very careful, he assured her: We do not want anyone to die during the study. She agreed to join.

He then ordered her to stop taking interferon, which was scary for her, explaining that he needed to get her white cell count up to 20,000. She had mixed feelings about stopping interferon: it had kept her white cell count under control, although it had done nothing for her Philadelphia chromosome percentage.

MID-NOVEMBER 1998

Judy Orem Makes a Journey

On the assumption that she had another six months to feel well enough to travel, Judy Orem planned a trip to New Zealand. She vacillated in her thoughts between a sadness that the end was nearing and a remote hope that she could enter the Phase I study at some point soon.

The weekend before she was due to leave for New Zealand with her family in mid-November, she woke up on a Sunday morning with a lump in her throat that was eventually diagnosed as an increase of her platelets from a normal range of 140,000 to 400,000 to over 1 million. The doctors were concerned that she might suffer a stroke in the near future. Brian Druker prescribed a pair of drugs to bring the platelets down; she would take them during her travels.

The trip went smoothly. Since she had stopped taking interferon in September, she was feeling better and better. Her mood and her memory improved.

Testing a Seriously Ill Patient

With the data in the trials showing that everyone was demonstrating improved white cell counts at the 250-milligram dosage level, Brian Druker felt it was time to test whether a really difficult patient could be helped by the drug.

That was why he tapped Sharon Godfrey for this task. Her white cell count had been hard to control with standard therapy. Then, when Druker took her off of hydroxyurea for just one week, she had gone from a white cell count of 5,000 to 125,000. It was clear that she was heading for a serious deterioration.

Druker told Sharon Godfrey that he had been testing a new drug that thus far had reduced patients' white cell counts at slightly lower dosages than he wanted to give to her, but that he had not yet treated a case as difficult as hers.

He wanted her to take the drug at the higher 250-milligram dosage and see what would happen. If it did not work, he would remove her white cells through a process called leukapheresis.

Druker admitted that he could not be sure that treating a more seriously ill patient with STI571 would not harm the person's health more if the compound did not work (because she was off of interferon and other therapies).

Sharon Godfrey Responds

A week after she went on the trial, Sharon Godfrey's white cell count stabilized at 125,000. Druker considered that an improvement over the dramatic increase she had experienced the previous week.

Judy Orem Enters the Trials

By mid-January 1999, with her white cell count up to 20,000, Judy Orem joined the Phase I trials—the ninth OHSU patient in the program—and began taking the drug at a 250-milligram dosage.

Several mornings a week she and four other patients—two at 200 milligrams, two at 250—arrived for blood tests at the OHSU bone marrow treatment room. Judy Orem and Sharon Godfrey, the two 250-milligram patients, were in quite difficult shape.

The pill gave Judy a "ravishing appetite." She joked to Brian Druker: "What did you put in that pill that gives us this appetite?" For so long she had not been enjoying the taste of food; but now "It was fun to eat again." She began putting on weight ("Most of us in the trials put on at least 20 pounds").

Judy Orem Gets a Response

Judy's white count came down in a few weeks. She was the first Phase I patient to get a white cell count below 3,000.

Judy's conquest of the platelet problem was icing on the cake. Here she had had this problem with the platelets, where the increase of her platelet count was a definite indication of disease acceleration; and then only six weeks later her platelets became normal. Judy Orem was the shining example of what a 250-milligram dosage did for a CML patient. (Platelets and white cells are not identical, but both are disease-related and the treatment helped both.)

EARLY FEBRUARY 1999

Normal White Cells for Sharon Godfrey

Four weeks later, Sharon's white count was normal. (In November 2001 it was still normal.)

For Brian Druker, Sharon Godfrey's improvement was a turning point. He knew then that there was something special about this drug.

After all, Sharon's white cell count had been difficult to control with hydroxyurea but after a month on this new drug, she was normal.

Indeed, both the experience of Sharon Godfrey and Judy Orem gave Brian Druker the feeling that something truly significant was taking place with the drug.

Here he had had two patients with an advanced phase of the disease, and the drug was controlling the disease with no serious side effects.

By this time the team began to realize that something good was happening. Patients' white cell counts were remaining normal. And a clear dose-response relationship was emerging, with a greater proportion of patients responding at higher starting doses. This was indirect—but very clear—evidence that the drug was responsible for the observed improvements.

The "diamond" appeared to be working.

These quick, positive results cannot be underestimated. It is extremely unusual for such results to show up so early.

In a normal Phase I trial, investigators enter patients with all kinds of cancers who had failed standard therapies—patients who would not have had a prayer of STI571 working on them because it specifically targets the Bcr-Abl oncogene that causes CML—and, as we learned later, only a few other rare forms of cancer.

All that investigators want to accomplish in these initial trials is to determine dosage levels that will not harm patients.

Nothing more.

But here were quick, reproducible decreases in white cell counts—clear-cut evidence that STI571 was working.

No one issued any press releases at the time. News of the drug's early success remained a tightly guarded secret among the team. They were scientists first and foremost. They had no idea whether STI571 would continue to work—and if it did, for how long.

They were witnessing history. They could wait to shout the news from rooftops.

And the news only got better.

Over time, the investigators increased dosages because the ultimate goal of the Phase I study was to determine what the maximum-tolerated dosage was. Once determined, Phase II patients would be treated at one level below maximum dosage.

The Little Old Lady

It was becoming clear to the Phase I investigators that they would have to reach dosages of 250 milligrams to attain maximal benefits. But then a serious problem occurred.

At the 250-milligram dose level, the investigators got a scare when one patient in Houston at that dose experienced a potentially serious problem. Because she weighed only 103 pounds—47 kilograms—and was elderly, the investigators called her the Little Old Lady. Blood tests of liver function had become abnormal.

Had they reached the highest possible dosage of STI571?

This was a distinct possibility because this same problem was encountered during animal testing.

The investigators knew they were at a critical juncture; in medical parlance they had perhaps arrived at the "maximum tolerated dose"—the dosage at which they were getting both efficacy but also toxicity.

And they were worried where all this was leading.

If additional patients on the trial were to experience similar or worse side effects, that could have meant the discontinuation of development, depending on the number of patients affected and the severity of the changes.

The Little Old Lady's setback sent shock waves through Novartis. Our clinical patient team was particularly concerned about the liver failure. We recommended that patients no longer be given 250-milligram dosages and that three more patients be enrolled at 140 milligrams. The investigators balked. A compromise resulted with three more patients enrolled at 200 milligrams. From then on, the patient cohort was increased from three to six at each dosage level.

Meanwhile, the Little Old Lady was taken off STI571. Her blood tests quickly normalized and two weeks later she was restarted on the compound at a reduced dose. Her liver problems never returned.

In addition, the number of patients treated at 200 milligrams daily (the next lower dose level) was increased to nine, as per the new arrangement. None of them developed liver changes.

Methodically, the investigators began increasing the dosages again, also hoping not to find other liver problems at higher dose levels.

The big question was whether the Little Old Lady was an omen of worse things to come—or simply an individual case?

As it turned out, no other patients exhibited liver troubles. The investigators took the dosages back up to 250 milligrams.

≋

Brian Druker would not let himself get overly excited: While it seemed that something very, very important was happening, he could still not be 100 percent sure. So he decided to stay quiet. He was not going to broadcast the news. Not yet at least. A bit superstitious, he was deeply worried that if he told someone, all the good that was coming from this drug would disappear.

Still, he could not believe the way patients were reacting to the trials. He knew that if the results held up, if other patients started reacting the same way, he might well have a breakthrough cancer drug on his hands.

He was still not sure, and for one very good reason: the lack of a cytogenetic response among any of the patients thus far. Getting such a response, of course, would be the true test of whether the drug could be put in the breakthrough category.

To Brian Druker, STI571 was proving itself to be a paradigm— even if there was no cytogenetic response from patients. But getting a cytogenetic response would be, in his phrase, a "home run." Attaining a white cell count reduction was one thing. It was a step forward for it *did* indicate an effect of the drug on the disease. But getting a cytogenetic response in a patient was the ultimate reward. Reducing the amount of Philadelphia chromosomes in the patient was the key to slowing down the disease and even from preventing it evolving to the terminal blastic phase.

To be sure, in the early phases of the patient trials, there was no way the investigators could have known if the drug had led to a cytogenetic response. During the first eight weeks of treatment with STI571, the investigators did not test the patients for changes in the number of cells positive for the Philadelphia chromosome. The bone marrow test that was required was considered too unpleasant.

That not a single patient in the trials had experienced a cytogenetic response in the first nine months of the study had proven disappointing to Druker and to the other investigators. He thought that perhaps the drug might simply render the cancerous cells harmless rather than eliminating them, in the same way that insulin did not cure diabetes, but made that disease livable with medication.

Druker told patients that with their white cell counts normal and with them feeling fine, they could live the rest of their lives like that. Many of these patients had not had normal white cell counts for years. But Druker remained disenchanted. He did not want simply to make CML a chronic disease; he wanted to eradicate it. After all, it was already possible to control white cell counts with hydroxyurea.

All that he had accomplished so far was helping to develop a fancy hydroxyurea. He clearly wanted more.

MARCH 1999

A Magic Moment

Then in March 1999 there was some dramatic news.

The investigators got their first indication that the drug might induce a cytogenetic response after all.

One of Charles Sawyers' patients in UCLA, who had been taking 300 milligrams of the drug, appeared to experience a major cytogenetic response, testing 50 percent Philadelphia chromosome negative.

The problem was that the patient had been tested for only 10 cells when usually a 20-cell test is called for.

It would take some time for the investigators to learn whether something truly interesting was going on.

APRIL 1999

Startling Data on My Desk

I was in my office at Novartis headquarters in Basel. It was April 1999. I had kept in close enough touch with our development staff to know that with STI we had a promising cancer drug in our pipeline.

However, I also knew that too many times after the first trials, we saw far too many failures to hold out much prospect for any drug at such an early stage of development.

And yet the data was astonishing. Of 31 patients who had participated in the Phase 1 trial: at 300-milligram doses, all had had a complete hematologic response and one-third had had a complete cytogenetic response (i.e., the disappearance of the Philadelphia chromosome).

All that I could think of were the thousands of patients who would want the drug immediately. And who could blame them? For just about every one of them, getting the drug might well be a matter of life and death.

A New Set of Challenges

I sensed trouble almost at once.

The bottom line was: At present we had no quick way to mass-produce this drug. Jörg Reinhardt would confirm this.

And even if we did decide to try to produce it quickly, it was certainly clear to me that we would come under tremendous pressure from both internal and outside groups who argued against rushing and taking risks for a drug that would produce such little return on investment.

I wanted to leave that argument aside. What preoccupied me was whether we could produce a drug in large quantities in the time required and how we could express our concern and care for the patients until we could deliver enough drugs.

We were good, and we had thousands of people working for us in far-flung places. And most important, we knew how to produce a drug in large quantities.

But it still took time.

There were procedures to follow. Governmental authorities had quality rules that had to be obeyed with utmost precision. We could not just add new facilities and hire new people overnight. For the most part we would have to rely on the employees we had in hand. As for our employees, they worked hard each day and then they went home to their families. They also took vacations. How much more effort could we expect of them?

And finally, there were our resources. We were able to maintain a very aggressive research and development program, and we had

ample financing to throw at a drug that we felt had a good chance of getting to market. We had finely tuned processes that allowed us to earmark resources, as we saw the need, as a drug showed signs of efficacy.

It was not our procedure to turn all our resources loose on a drug at the very beginning of the development process.

That was too risk-laden. That did not seem prudent. We knew how to proceed: slowly, prudently, cautiously—one step at a time.

Then along came these results.

As I looked at that piece of paper, I began to sense that this drug could test us in ways that we had never experienced before.

My mind began racing in a thousand directions.

How could we gear up our resources to produce commercial quantities of the new drug quickly enough to save as many lives as possible?

What would be a fair price for the drug? And how could we guarantee fair distribution among patients?

Would it be possible to persuade the authorities, in the United States, Europe and elsewhere, that they must help get the drug to the public quickly by speeding up their approval processes?

And what should one say to the public about this drug? If we overhyped it we would mislead numerous patients suddenly imbuing them with newfound hope while risking huge embarrassment to ourselves. But underplaying the significance of the drug could give others the impression that we ourselves had doubts about its value.

Too Overwhelming to Ignore

We would have to find a way around the constraints. The quality of the drug was simply too overwhelming. A moment like this one comes perhaps once in a lifetime. It made no sense to miss out by being overly conservative.

And so, at the senior levels of Novartis we implemented what I would describe as "innovation management."

Innovation management assumes that the private sector is the most capable of carrying out innovative drug discoveries.

Innovation management then raises these questions:

Which compounds are likely to provide a substantive benefit to the patients and favorable returns for the company?

How do we structure the development process to optimally manage speed, quality, and costs?

What organizational structure optimally supports the integration of research, development, marketing, production, and customer insights?

Creating the right kind of climate is a big part of innovation management: a climate that allows risk-taking; where employees feel motivated and have a sense of ownership of the product; a climate that encourages speed as well as good planning.

The organization will have many things to do once a drug shows early potential. Each aspect of the project must be carried out quickly and carefully. The organization must collaborate; it must regard the new compound like a new common baby.

In the end, the CEO must rely upon colleagues to do their jobs. To guarantee that that they will, competent people are essential.

The CEO's most important contribution is to push the process forward when it needs a push or to apply the brakes if needed. That does not mean the CEO gets involved in every single decision—not at all. But at times it may be imperative for someone at the top to place a phone call, or to suggest to a colleague to make a phone call. The CEO has to know when to make the suggestion, and when not. There are no rulebooks for this kind of thing. One just has to sense when to get involved, depending on the project, the kind of intervention, and an understanding of how the organization will perceive such intervention. Of course it is crucial that, in the end, the organization is aligned along common priorities.

Minimizing the Risk

Routinely, our strategy at Novartis is to develop systems that tend to minimize the risk throughout the drug development process. For instance, we typically spread the costs of developing and producing the drug over time, so that we can monitor the progress of the drug every step of the way and only produce the needed quantities. We conduct small patient trials before we engage in large ones.

I did not think this was the right strategy for STI571.

I sensed that the risk was high. Though my goal would be to get the drug to people before they died, if the drug was stillborn, critics would hardly give me credit for trying.

They would simply say, "He should have known better. He wasted the company's resources. He wasted his employees' energy. He wasted time and has forgone better opportunities."

Such high-stakes gambling the first time around would eventually be forgiven, but if it were ever repeated, people would start to doubt my capabilities. I was all too aware of these risks.

But I deeply believe that if you want a climate in a company where people take calculated, but not unreasonable risks, you have to demonstrate that you are not afraid of taking risks yourself—and, most important, you are not afraid of failing.

Knowing the risks, I might have been expected to play a much more hands-on role than I did. But I had complete confidence in my senior colleagues, particularly Jörg Reinhardt, global head of development, and Andreas Rummelt, head of global technical operations, as well as Alex Matter, the head of oncology discovery research, and Greg Burke, the global head of clinical development in oncology.

I saw myself, not as someone who stood over these people, but more like a midwife who watches over the birth of a baby, gives encouragement and support and only intervenes when necessary, making sure that everything goes well.

I knew that my senior colleagues would mention the huge upfront costs that would be involved, wondering if we really wanted to spend that much money so quickly for a compound with a small market potential.

Being in close touch with Jörg Reinhardt was unusual on a specific project. The system has been designed at Novartis so that we would not ordinarily have to discuss development issues outside our innovation management board meeting. I would only hear from him if something went wrong or exceedingly well.

But so many things surrounding this drug were different, most important of which was how much faster the development process would have to be than in normal circumstances. All of this was highly risk-laden.

This is where the culture at Novartis comes into play. It is a culture that has learned to grapple with high risk by giving our researchers as much freedom as possible without losing focus and alignment. In the past, before the merger of Ciba-Geigy and Sandoz, some of our scientists complained that they were suffering from a lack of freedom. After the merger, we focused on creating the right working atmosphere. Hearing Alex Matter say that he doubted that this drug could have been produced under the old, more conservative regime was a sign that we had succeeded. We have gone in for high-risk projects, and the scientists naturally react positively to that attitude.

We also have tried to make speed part of our culture, but not the kind of speed that turns into haste. We encourage work that is qualitatively high—the first time around. That is why we have always placed such importance on getting any trials right the first time.

Gleevec did not come out of the blue. There is a whole culture at Novartis that puts a focus on the discovery arena. We have been concentrating our efforts for some time on high-risk research, on truly innovative science where the chances of success may be low but a certain investment is still required. Our goal is to benefit the human race in a unique way; we want to be true innovators, bringing new, important drugs to our patients, drugs that cure and prevent diseases, drugs that improve the quality of life.

Given that background, it becomes easier, when a product like Gleevec pops up, for the company to rearrange its priorities and to make sure that the product gets the proper attention.

Our company's culture means that employees will make an extra effort to produce a drug more quickly, not because someone said there is more money in it for them, but because they understand that what they are doing is of great benefit to patients.

As the development process moved forward, I thought of my sister, Ursula. The doctors had not been able to save her then; years later, thanks to new drugs, they would have been able to keep her alive. The connection between Ursula's tragic death years ago and the data on the paper in front of me seemed crystal clear. suddenly I could visualize thousands of Ursulas out there today, suffering from CML, and that made the decision to accelerate the production process of Gleevec much easier.

We faced a number of challenges but the two largest were:

The drug might not pan out.

And, we might be too slow in getting the drug to the market, turning patients into activists who would lobby us aggressively.

This was, after all, the age of the Internet. News of the remarkable qualities of the drug would spread instantly. Demand for the drug would be instantaneous as well.

Our task at Novartis was to move to industrial-scale production of Gleevec at once. To contemplate such a step—with data based only on Phase I trials—appeared quite hasty to some of my colleagues.

We needed more time to find out whether the Phase I results would hold up in the long term, they said. We should wait to determine whether there are new side effects.

They were being cautious. I respected their prudence. But I did not want to wait.

Typically for Phase I trials, a pharmaceutical company would only produce the drug at "lab scale." Treating only 30–40 patients requires just a small amount of the drug and that was about all that had been produced so far.

A Dinner for 20,000

To go from a small amount to producing tons of a drug is no easy feat. Here's why:

Think of making a dinner for two people.

Then think of making a Thanksgiving dinner for 20 people—certainly more complex, but manageable.

But what if you had to jump from dinner for two to dinner for 20,000?

It might make sense to practice on some number in between 20 and 20,000.

If we were going to scale up production for Gleevec with an accelerated schedule, what we would have to do was to jump right to huge tonnage—to having 20,000 people over for dinner.

To wait until the stage of the Phase III trials to scale up production—as is the usual practice at Novartis and most companies—would have meant significantly limiting enrollment into our clinical trials and perhaps getting FDA approval with no supply of the drug to speak of.

If demand for Gleevec was going to grow as quickly as we suspected, the only choice would be to establish a lottery system for those who wanted the drug.

The trouble with a lottery system is that some people win, but some people lose.

I did not want to reach the stage where we had to resort to lotteries. I did not want huge numbers of patients stalking us at every turn, demanding that we stop lagging behind in producing this drug.

My Business Is Risk

Jörg Reinhardt called a meeting of the handful of people who were running the Phase I patient trials. For him to call such a meeting was unusual. Reinhardt has 5,000 people working for him in the development program. Normally he would meet with the leader of a drug project about four times a year. Otherwise, he read reports.

But this time, the coordinators of the STI571 Phase I trials made sure to keep Reinhardt up to date.

Until the April 1999 data arrived on my desk, Jörg knew only that ten or so patients had responded very well early in the trials. Such small numbers, however, were not enough to persuade him, as he had seen too many failures. The data that arrived on my desk served as a real wake-up call for him. He was impressed.

Jörg Reinhardt was not especially concerned with the risk factor. He always had risks in his projects: "Most of what I work with every day is failure," he likes to say. "My business is risk in development. Most of my projects die."

Still, he had a hunch that this drug was going to succeed. He had developed a gut feeling over the years and his gut was telling him Gleevec was a winner.

Asking the Impossible

He knew full well that nothing was certain at this stage. A severe side effect could develop in one out of 1,000 patients and that would be the end of the trial. And these side effects might arise only at the very end of the trial. It could take two or three years. Nothing was certain.

But this time he had a hunch.

"Look, this is probably something spectacular," he told the patient trials team. "Let's discuss how we can accelerate the development process. How can we make the drug available to patients in a year."

The faces of the people in the room said it all.

What Jörg Reinhardt was asking seemed to be impossible.

If we at Novartis could file for approval with the various authorities at the end of 2001, that kind of schedule would seem most reasonable. But Reinhardt was asking that the drug be filed by December 2000 and that it be on the market by the middle of 2001.

He wanted the drug on the market in two years!

That would take nothing short of a miracle. So it seemed at the time.

The people in the room had all sorts of opposition to Reinhardt's schedule.

They argued that the Food and Drug Administration still did not know about the spectacular results of the Gleevec patient trials. How could we be sure that the FDA would allow us to make a submission based on significantly less data than we would have by the end of 2001?

Given the accelerated schedule that Reinhardt was talking about, at best we could only report on Phase I and Phase II trials. The FDA would have to accept a submission that did not include Phase III results.

(In February 1999, the investigators began laying out what the Phase II patient trials should look like. Two months later, they began to prepare for Phase II in earnest.)

People in the room thought we were fooling ourselves.

The skepticism of our colleagues was understandable. Novartis had never before accelerated the development and production of a drug to such a degree. It normally took between two and three years to produce such drug quantities. Occasionally, we had shaved off a few months, but that was nothing compared to what was being asked of the people in the room.

We wanted to knock an entire year off the development and production schedule!

Jörg Reinhardt was challenging Novartis to have the drug ready for filing by December 2000—in another 20 months.

Jörg should have been worried over whether the production team could do the job in such little time. But something else bothered him much more: the way patients would continue to respond to the drug.

There were so many things that could still go wrong.

Not all the toxicity data was in.

Nor was it known whether the capsule was stable for a long time;

if the capsule needed to be reformulated, new patient trials would have to occur to test the new capsule.

Nor could anyone guarantee that the side effects of Gleevec would remain tolerable. But we had other things to worry about, mostly how could we speed up the development of STI571.

Usually, Novartis produces a drug first in a pilot plant in Basel, and then in a chemical production plant, also in Basel. Because everyone knows and is comfortable with one another on the development and production teams in Basel, the hand-over from development to production always goes smoothly.

With Gleevec, Andreas Rummelt realized that he did not have the usual luxury of two or three years to produce a drug in sizeable quantities.

When there is a demand for truly large quantities of a drug, production shifts either to our production facilities at Ringaskiddy, Ireland or at Grimsby in the United Kingdom.

Giving Ringaskiddy a Chance

It was decided together with Hansjürg Wetter, the head of chemical operations, to move directly from the development pilot plant in Basel to the major production site at Ringaskiddy.

Jörg Reinhardt noted: "We all sensed that if and when it became clear that Gleevec worked on forms of cancer other than CML, there might be an uncontrollable demand from doctors and patients. So I asked Andreas Rummelt to try to act very boldly and produce the maximum amount in as short a time as possible. Hence, we skipped the Basel production step, the first time we ever did that. As a result, we had to send certain people from technical development to Ringaskiddy, while in the past they would simply go from one building to another in Basel.

"We knew that we were taking a certain chance in skipping the Basel step because a good deal depended on how well the development people in Basel would get along with the Ringaskiddy folks.

Development and production people don't always have the same work styles, and we knew it would take some adjustments on both sides. But it worked."

Reinhardt, Rummelt and their staffs set a deadline of June 2000 for one ton of Gleevec to be produced. That was just over a year away.

Had they followed the usual procedure, they would have needed at least two years, perhaps more.

Andreas Rummelt, excited by the challenge, knowing that we were not going to let money stand in the way of accelerating the production of Gleevec, worried very little about getting approvals for capital investment. He simply let us know what he planned to do— and he got around to the paperwork later. He enjoyed the freedom of maneuver: "It was exciting to prove that you can not only think out of the box, but that there are possibilities to be innovative, to be quicker."

The only trouble with the original plan of producing a ton of Gleevec in a year was that it overlooked the more immediate needs of patients in Phase I trials and of patients who would be starting Phase II trials.

Therefore, Andreas Rummelt and Greg Burke, global head of the oncology clinical development unit, agreed on the following plan:

1. A small amount of the drug—some 50 kilograms—would be produced by September 1999. That would provide a supply for patients still in the Phase I trials and for patients starting the Phase II trials.

2. Another 500 kilograms would be produced by January 2000 and 600 kilograms in May.

3. A further 1400 kilograms would be produced by August 2000 (including milling and blending in Stein, Switzerland).

4. Another 23 tons would be produced in 2001.

More Good News

In June 1999, Charles Sawyers was getting ready to board a plane to attend a meeting in France. A fax arrived. He looked at it and could not believe his eyes. It reported that one of his patients in the Phase I trials, Ellen Froyd, of Carpinteria, California, had showed a complete cytogenetic response. He called the lab and asked the technician to pull out Ellen Froyd's records "and tell me that's true." The technician got the records and assured Sawyers that it was indeed true. Sawyers called Ellen Froyd, and then placed a call to Brian Druker, but Druker had already left for the meeting in France. Sawyers was only able to inform Druker of the news about Ellen Froyd as Druker got up to speak at the meeting. Druker's first words were: "I've just received some wonderful news from Charles Sawyers . . ."

For Sawyers, Ellen Froyd's remarkable improvement was a turning point in his thinking. He and his colleagues were used to interferon taking a year or more before any kind of cytogenetic response occurred. But here was Ellen Froyd on Gleevec for only five months, obtaining a complete cytogenetic response. That was, Sawyers told himself, incredibly fast.

No cancer treatment had been shown to attack the core of CML in the patient's bone marrow with such tolerable side effects.

Hydroxyurea lowers white cell counts but has almost no effect on the Philadelphia chromosome.

Interferon can reduce the percentage of cells showing the Philadelphia chromosome in the bone marrow but the side effects are often horrendous.

Sawyers was now convinced that Gleevec was something spectacular. "We have now beat interferon for sure," he said to himself gleefully. "I now knew that not only does the drug work, but it's far better than we could have hoped."

For the next 54 patients who were enrolled in the trial at daily doses of 300 milligrams or more, the results were truly astounding: all but one entered a complete hematologic response, which typically occurs in the first four weeks of therapy.

JULY 1999

The Focus Is on Ringaskiddy

Once the decision was made to speed up production of the active substance in Ringaskiddy and of the dosage form in Stein, Switzerland—to provide tons of Gleevec active substance and millions of capsules, instead of just kilograms and thousands of capsules—the whole company seemed to light up with enthusiasm. People started to work day and night.

The focus was on the major drug substance production site in Ringaskiddy, Ireland.

On July 28, 1999, Maeve Devlin, head of production in buildings 552 and 553 at Ringaskiddy—the buildings where Gleevec was produced—began hearing that Ringaskiddy might be requested to take on an entire launch of a drug.

It was then that she and the others in the managing team at Ringaskiddy learned that Gleevec was in patient trials and that the trials were going so well that Novartis was ready to choose a site for making the drug in commercial quantities.

The actual work at Ringaskiddy would begin in September.

The entire Ringaskiddy facility had 319 employees, 93 of whom worked in the buildings where Gleevec would be made. An air of excitement permeated the Ringaskiddy plant. It had been producing drugs for Novartis since 1994, but it had never been selected as a launch facility, the place where the process of first commercial production of the drug substance actually begins.

Until Gleevec, the Ringaskiddy facilities only came into the picture after a drug had already been launched at a Novartis production site in Basel. Ringaskiddy came into play only if Novartis wanted to make larger batches of the drug. All the teething problems were ironed out in Basel. Before Gleevec, the Ringaskiddy team in buildings 552 and 553 had been making material for a half dozen drugs, including two top-selling ones, terbinafine that fought toenail infections; and the cholesterol-lowering drug, fluvastatin. Ringaskiddy had never been the only production site to make life-saving drugs.

≋

Beginning from Scratch

Now, with Gleevec, Ringaskiddy was going to get a chance to take on a launch from scratch.

Typically, a "prototype adequate" campaign would be run at the pilot plant in Basel to verify that the manufacturing process was stable. That campaign routinely took six months.

We decided—in the interest of saving time—to go straight to the launch campaign, bypassing the "prototype adequate" one, the first time we did this at Novartis.

We looked for ways to save more time. And we hoped that the Ringaskiddy team would provide some answers.

The responsibility fell heavily on the shoulders of the Ringaskiddy managers. Jerh Collins, plant manager in building 552 and 553, felt that new responsibility: "Time was against us. From the early stages, in September 1999, we had to get the product up and running by November of that year. We really needed six months, not two."

The managers received relatively little information at the start, only that the drug had to be fast-tracked.

To motivate employees, Ringaskiddy managing director Franz Sutter assembled his staff and essentially gave employees a pep talk: He noted that Ringaskiddy was entering a very exciting phase. By being selected for the launch, the plant would be getting into manufacturing at a much earlier stage than had been the practice. He told them that if everyone did a good job, it would help secure their future. He stressed that they had the great opportunity to make something that had the potential of saving lives.

And they would be the only people making the drug. Jerh Collins remembered Sutter's words as "very powerful, very powerful. It was the first time that people felt that they were really, really close to our customers, and they started thinking of people who in their own families had critical ailments. It brought the message very close to home."

SUMMER 1999

Phase II Trials Start

While we were gearing up the manufacturing process of the drug, we also decided to begin Phase II patient trials. Their purpose was to validate the results of the Phase I experience.

Novartis's John Ford and the three investigators faced a true ethical dilemma at this stage.

Now that it seemed clear that the drug was really having an effect at higher doses, what should they do about patients who had had the misfortune of entering the study at low, ineffective doses?

There was literally no one to whom they could turn for guidance.

This has rarely, if ever, happened before in a Phase I cancer study, for good reason: First, the level of activity is often quite marginal and is rarely indicative of a convincing effect on the disease process; and second, usually by the time this point is reached in a Phase I trial, the first cohorts of patients are either dead or no longer fit enough for retreatment.

It turned out that a solution was not that hard to decide upon.

The team simply re-entered patients into the trial at what they believed to be effective doses.

Accordingly, the very first patient who was treated originally at

25 milligrams was put back on therapy on April 26, 1999, at 300 milligrams daily.

Virtually at once, his white cell count normalized and remained that way.

(In time, some 10–15 percent of the STI571 patients in the Phase I trial achieved a complete response at all doses above 300 milligrams. That figure appears to be increasing steadily with increased time on drug; indeed, in August 2001 it was about 25 percent.)

Getting those good results at the higher doses in chronic patients, John Ford and the three investigators decided in the spring of 1999 to expand the Phase I trial to include 59 CML patients in the blastic (final) phase of the disease.

The investigators were skeptical that these, the most serious CML patients, would respond to Gleevec. One indication of their skepticism: they allowed only three blast crisis patients to be enrolled in the Phase I effort at first. Surprisingly, within a matter of a few days, all three patients responded to the drug: Their white cell counts came under control and the percentage of blasts dropped. Charles Sawyers remembers the shock that came over him: "Normally, these patients have a month or two to live. All you can do for them is to give them morphine, and tell them you're sorry."

Over the next two to six months, the three patients relapsed and died. To Sawyers, their relapses were equally shocking. "One week they were fine, another week their leukemia is back." He was clearly disappointed.

In the summer of 2001, Sawyers and others published a paper suggesting the reason for their relapse: either mutations had occurred in the Bcr-Abl kinase that prevented Gleevec from binding to the targeted molecule; or the Bcr-Abl gene amplified in the leukemia cells so that the number of targeted molecules "out-competed" the drug.

Sawyers thought the next step was to figure out what would be the right combination of Gleevec and other drugs to help blast phase patients.

Sawyers was hopeful. After all, he had never imagined that Gleevec could offer any help to blast phase patients: "I am a very conservative person. But when those blast phase patients responded to Gleevec, that was like Lazarus rising from the dead. And believe me, I am normally a very skeptical person."

≋

The investigators were so pleased with the emerging results of the Phase I trial that they were already considering new trials on the efficacy of STI571 in patients with solid tumors. But, without enough of the drug in supply, they were forced to postpone the start of those trials for a full year.

Once the word began to spread that STI571 might well be a breakthrough drug, we at Novartis would undoubtedly face pressure to produce the drug in much larger quantities. A seemingly simple decision, it would prove to be far more complex than any of us might have thought.

≋

The Phase I patient trials for STI571 exceeded expectations. Before they began, the scientists believed they were on to something, but no one could be sure until a significant number of patients showed improvement from the drug. The trials encouraged the scientists to move quickly to verify the original findings; but before those scientists got too far, word began to filter out to CML patients who wanted nothing more than to find a drug like STI571.

5

The Petition

As word began to filter down from the patient trials that Gleevec was a breakthrough drug, patients began clamoring for it and so did research scientists.

In July 1999, Professor John M. Goldman of the Hammersmith Hospital in London, along with 30 other scientists from other parts of Europe, listened to Brian Druker talk about the wonderful Phase I patient trial results at a meeting in Bordeaux, France. Druker was able to report for the first time on the remarkable reactions he and his investigators had been getting from some 15 CML blast patients who had entered the Phase I trials the previous April.

Goldman found the data very exciting. He called the results a dramatic contribution to the management of leukemia because of the molecular link-up. That the drug was easy to administer, and appeared to be nontoxic was appealing as well.

Goldman and the other scientists asked Novartis's John Ford when could they obtain the drug for research purposes.

It would take another 18 months, Ford replied.

The scientists' faces dropped.

Ford insisted that it was a difficult drug to make, requiring numerous chemical processes. Nor was it a high priority for Novartis, given the small number of CML patients.

Perhaps, if production is indeed accelerated, the drug could be available by autumn 2000, Ford suggested. But even that piece of news did not mollify the horrified scientists.

After returning to England, Goldman wrote a letter to Novartis, trying to speed things up.

Brian Druker was also acutely aware of the problems arising from the Gleevec shortage. It hit home to him when he appeared at a conference in Biarritz. Sitting in the audience was Peter Rowbotham, whose wife was a CML patient, and who had been attending conferences and reporting his findings on the Internet for some time. To Druker, Rowbotham's presence at the conference reflected the mounting public pressure on him to provide the drug to CML patients. Here was Druker presenting findings of the Phase I trials to a small group, seemingly well out of hearing of the public and the media; yet Peter Rowbotham was bound to report back what he heard at the conference to CML patients via the Internet and those patients would in turn send e-mails to Druker, asking to get into patient trials. Druker had only two slots a month for patient trials. And it was still not clear to Druker that we at Novartis would be able to produce the drug in sufficient quantities for the wider CML patient group.

We were certainly doing all that we possibly could do.

But, for Brian Druker, all of this was proving quite frustrating. He had been invited to provide the first public glimpse into the Phase I patient trials on Gleevec the following December 1999 at a major conference in New Orleans. He knew how spectacular those results were. He worried that if we at Novartis could not provide the drug quickly and if there were no further patient trials in progress over the next few months, his presentation in December could raise false expectations among countless numbers of doctors and patients.

≡

In the fall of 1999 production was moving swiftly—though not swiftly enough for a public slowly learning about the great results we had obtained in patient trials.

The 12-step manufacturing process of STI571 included seven isolated steps as well as 11 other chemical steps—an isolated step may have two chemical reactions.

Over the next two months—September to October 1999—the Ringaskiddy plant took in all the raw materials and focused on seven of the 12 chemical steps in the manufacturing process before turning over the substance to Stein, where it was put into capsule form. Jerh Collins, the plant manager at Ringaskiddy, explained: "The key raw materials comes to us as a powder. We add our starting materials. It's a bit like making a cake: You add your flour, milk, raisins and currants, dried fruit, etc. They are your starting materials. You mix it, heat it up, and then crystallize it by cooling the material down or adding further solvent. You then get a powder coming out of a solution which you filter off." The Ringaskiddy team had to get the first step under way by November 1999, and it made that deadline.

How did the Ringaskiddy employees move so swiftly? They improvised.

One issue had to be solved quickly: some of the materials required special handling precautions. But nothing existed at the plant that could protect the employees from the potential harm that could come from handling the materials. An American company that made space suits for NASA provided the necessary protection to work so employees could unload the materials without risking exposure. The solution was a containment device called a "doverpac" that had its own unique docking system plus temporary containment suits developed with local engineering partners.

Each of the seven isolated steps had to be completed in one month. There were five to ten batches of materials associated with each step. Jerh Collins recalled that this work "took tremendous effort" with many people working evenings and weekends.

Robert Doyle, a quality control group leader, recalled some of the difficulties that cropped up. Instructions on how to do testing came from Basel in German; they had to be translated quickly. There was much pressure to get data analysis done quickly: " Before I could finish, others at the plant were ready for the first step of production. In my department, we had four people working on the testing of STI571. We couldn't afford any hold-ups in productions."

Six steps remained to be done—steps that would normally have taken a year. But we did not give Ringaskiddy a year. We set a target of June 2000—seven months away—and of August 2000 for the Stein facility to do the milling and blending.

≣

Moving ahead as fast as we could, we continued to come under more and more public pressure to produce the drug in larger quantities as quickly as possible. The crunch came that fall of 1999 and it focused on a Canadian CML patient with much grit and much heart.

In Gleevec circles, Suzan McNamara will always be known as the woman who began the petition to get the drug distributed more quickly. She led a relentless campaign, organizing CML patients to press Novartis to assure the widest and swiftest distribution of Gleevec possible.

Suzan McNamara, born in Montreal on November 2, 1966, was diagnosed with CML in March 1, 1998. During the six months prior to the diagnosis, she had not been feeling well. She felt her energy draining. She felt tired. During the nights she experienced night sweats and in the mornings she woke up feeling ill. She was losing weight.

On March 1, 1998 she noticed that her spleen had become enlarged. The word "spleen" kept appearing in the medical text that she consulted from time to time. More alarms. She went for a blood test the next morning.

What's Wrong?

Twenty-four hours later—she remembered it was a Tuesday morning—Suzan showed up at work only to receive a phone call from her doctor at 8:30 a.m.

"We feel you could have a form of leukemia."

"What's leukemia?" she asked. She knew very well what leukemia was but her mind had gone blank. No longer calm, she asked, "Blood cancer?"

"Yes," the doctor said quietly, "we consider it a form of blood cancer, but we need to do more testing."

Arriving home, she looked through a ten-year-old medical text for information about CML. According to the outdated book, she had three to five years to live. She dropped to the ground upon reading that horrific news.

She felt as if her spirit had just left her body. Thus began her long depression. She wanted to live; yet no doctor could tell her that she would definitely survive the disease.

I Want to Live

When Suzan saw an oncologist for the first time, he told her that she could live three to five years with hydroxyurea; but with interferon, or a bone marrow transplant with a suitable donor, she could live longer.

Suzan had a white cell count of 380,000—far higher than normal. The doctors wanted to get her white cell count down before she slipped into the next phase of the disease. So she went on hydroxyurea. While the drug has no side effects, the patient can feel weak since good cells are being killed along with the bad.

Suzan remained in the hospital for a week. On her third day, she began daily injections of interferon.

Returning home, she became quite ill and weak from the interferon. Work was out of the question. She lost 20 pounds in a week. She was too frightened to look at herself in a mirror.

Her bosses arranged for her to work from home, but she found it too difficult to concentrate. She trained someone to replace her and went on disability leave. She was 31 years old.

Suzan vowed to fight the disease aggressively. She turned to meditation, yoga, acupressure (like acupuncture but without the needles), and massage therapy, as well as organic foods.

The doctor doubled her dose of interferon; while most patients received at most 9 million units per day, she now received up to 20 million units daily. She remained depressed and frightened.

Still, the interferon and healthy lifestyle was working. From having 100 percent cancerous cells at the time of her diagnosis, within a year she dropped to 35 percent. In the initial six months she began feeling stronger and came out of her depression. But the interferon had taken its toll: her hair fell out. She had a severe rash on her face and the toxicity from the drug made her feel sick 24 hours a day.

Six months into her interferon treatment, doctors found a match for her for a bone marrow transplant. She doubted that she was either mentally or physically able to go through that experience. She knew that the surgery killed off one's immune system. The odds that only 50 or 60 percent of those who had the transplant would have a long-term life scared the hell out of her. She said no to the surgery; but the doctors kept pushing her.

What depressed her beyond anything else was the doctors' saying that she would have to stay on interferon for the rest of her life. She occasionally asked herself if maybe death was better.

Surfing the Internet

In the fall of 1998, surfing the Internet for information about CML, Suzan latched on to a CML support group chat site on the Internet that was part of Egroups.com (the site is now part of Yahoo! groups).

Patient trials for STI571 had begun at the three medical centers in the United States, but at this stage the positive results had not been made public. Had Suzan known of those results, she might have begun her campaign to get her hands on Gleevec even sooner.

Eager to learn how others with her disease were coping, she

typed in her name and signed up for the site. It had 400 patients on its list. The Internet support group helped her out of her depression. But at first, her search for information proved outright disheartening. Every sentence she read seemed to contain the words "death rate," "fatal," "percentages." Even the material about clinical trials for experimental drugs brought tears to her eyes.

It was not until April of 1999 that Suzan began hearing about a new drug called STI571 from the same Peter Rowbotham who had been at the Biarritz conference. He had become something of a saint for CML patients. His wife was a CML patient who, when interferon stopped working for her, turned to the new drug.

Peter began to spread the word about Gleevec to everyone. Attending conferences, gathering information wherever he could find it, he acquired a reputation as a layman who could turn arcane medical language into straightforward prose and make it clear to the support group.

When Suzan McNamara read Peter's comments about STI571, she thought it sounded good: "When you're in my situation, you love hearing about something like this."

She read of a few people within the support group, CML patients who were failing and for whom no other drugs were working, who had begun to take the drug. Thanks to the capsule, they had regained their strength and their weight; their white cell counts were lower than when taking interferon; and they were experiencing few or no side effects.

And, best of all, CML patients who were taking the drug were reporting to the support group that their quarterly bone marrow tests were showing a decrease in the percentage of cancer cells.

A Dream Come True

Each morning Suzan rose enthusiastically and went straight to her computer. The news of the new drug was literally giving her a bounce when she woke up.

Fast forward to August 1999: Suzan's condition had been stable for the previous six months. She was not trying to get into an STI571 trial. Because interferon had produced a cytogenetic response (35 percent Philadelphia chromosome positive over the last six months), she thought it only fair that less-well patients get a shot at the drug first. And, at any rate, she continued to feel good.

Then, suddenly her white cell count began rising, no longer controlled by the interferon. And the percentage of blasts—or immature white cells—increased from 2 percent when she was diagnosed (disappearing under interferon) to 5 percent. A CML patient moves into the accelerated phase with 15 percent blasts, and into the blast (final) phase with over 30 percent.

Doctors increased Suzan's dosage of interferon several times. She began feeling unwell again. She began taking Ara-C, getting two daily injections. Over the next ten days, the effects of the drug sent her to bed semiconscious and terribly ill.

Her thoughts finally turned to STI571. She felt that, now that she was showing signs of acceleration, she could and had to get on the trial.

Her physician in Montreal tried to get her into a clinical trial in Houston but was told that they were not accepting any more patients in the chronic phase. Houston would take her as an accelerated phase patient if and when she rose to 15 percent blasts. Houston had new trials for chronic CML patients scheduled for March 2000. Suzan felt the same jolt that she experienced at the time of her diagnosis but this time, rather than weakness, she sensed a new power within her that would enable her to fight. The first thing she had to do was to get on the Gleevec clinical trial and get off of interferon. Interferon was not keeping her white count down; it was just making her sicker, weaker, and not giving her body any strength to heal.

Small Supply

In mid-September, 1999, Suzan sent an e-mail to Brian Druker and told him her story. He phoned her to say that he wanted to help her, but right now there was hardly any supply of the drug. She would have to wait another three to four months perhaps, before he could start another clinical trial.

The news was depressing. Druker impressed on Suzan that he was trying his best at his end to make Novartis aware of how wonderful the drug was, to get their attention and to give him more drug, "but maybe a patient can do something, coming from a different angle."

"Okay, I will try to think of something to do," Suzan said. While in bed that evening, she recalled visiting a petition site on the Web: *petitionpetition.com.* The site allows people to compose and distribute a petition on any subject. Those who want to sign a certain petition only have to log in, type a name, occupation, and age—that counts as their "signature."

The next day she rose early, e-mailed the support group and asked them what they thought of this idea. The replies were overwhelmingly positive with offers of assistance from each e-mailer. Suzan felt that at most her petition would obtain 200 signatures. Taking no more than half an hour, she copied material that Peter Rowbotham had gathered for the support group site and pasted it on to her new petition.

On September 21, 1999 she sent the petition out. The first day 50 signatures came in; then it snowballed to 100 each day. By October 11—just three weeks later—the number of signatures had reached 3,030.

The next day, collaborating with Peter Rowbotham in getting a polished draft, Suzan sent the letter and the signatures off to Novartis via Federal Express. The letter was addressed to me and read in part:

> . . . Many of us who have signed the petition believe that the theory behind this new drug is fundamentally sound, and the specific targeting of the Bcr-Abl oncogene by STI571 presents an unusually good opportunity to increase the survival of CML patients significantly. It is not impossible, based on the results to date, that the new drug will prove to be a functional cure for some patients.
>
> Because of the particularly good prospects for this new

drug, we have viewed with growing concern our belief (based on information from various sources) that the supply of the drug has not been sufficient to expand the trials as fast as the evidence to date would warrant. While we are not experts, we are convinced that through the efforts of Novartis the present trials could be expanded significantly, and we anticipate that this would lead to a quicker certification of STI571 for general use. It is imperative from the perspective of CML patients that these trials be advanced with all possible speed, commensurate with the responsibilities of the investigators.

We therefore ask for your assurance that everything will be done to produce a sufficient supply of STI571 to ensure that the trial investigators are not held up in any way at all in trialing this new drug, and in advancing to the certification that we anticipate.

The letter could not be ignored.

In the case of Gleevec, one important role I played as the Chairman and CEO of Novartis—part of what I call innovation management—was to impose some order on a part of the process that could have easily gotten wildly out of hand.

This was when people clamored to have access to the drug. If we had allowed just anyone to have access to the drug, chaos would have followed. We had to establish criteria regulating who could gain access to the drug in the early stages, and who could not. Otherwise, our patient trials would have been tainted, and we would not have been able to move forward as quickly as we did in winning FDA approval and in getting the drug to thousands of patients. We had to resist giving the drug out unconditionally. The trials had to be conducted according to the highest standards.

This was by no means easy.

With the petition that Suzan McNamara had sent us and other appeals that were arriving, we appeared to be at the very beginning of the "patient revolt" that I had warned colleagues about. I wanted us to deal with the issue as quickly and as fairly as we could.

As word began to spread about the beneficial qualities of Gleevec on the Internet and elsewhere, letters and e-mail began streaming in to Novartis from patients and friends of patients, and relatives of patients—all wanting access to the drug. The appeals came from celebrities and political figures; and from the wider public as well.

Meanwhile doctors and patients kept calling us, seeking the drug.

We sensed that Suzan McNamara represented a large number of CML patients, and we wanted to assure them, through her, that we were on the case.

≋

Just two and a half weeks after Suzan's petition was sent, on November 2—Suzan's 33rd birthday—Brian Druker phoned Suzan to say that her petition had really stirred things up.

"You're kidding?"

"No, I'm not kidding. I have heard that Novartis is responding to you as we speak. Things are looking good. That's all I can tell you. You can't tell the support group. You have to keep it a secret until you receive the letter." Not being sure what the Novartis letter would say, Druker wanted to be cautious until it arrived.

Five days later—on November 7—the Novartis letter, signed by James S. Shannon, head of Clinical Research and Development, arrived at her home and it read in part:

> Although the preliminary data is encouraging, STI571 is still in very early stages of development and its safety and efficacy profile, as well as its optimal dosing, are currently being evaluated in three clinical trials in the United States and Europe. Conscious of the impact this agent could have on patients with CML, Novartis has placed a very high priority behind

expediting the compound's development with all diligence. Additional resources have been devoted specifically to substantially expand the drug production capacity for STI 571 in order to accelerate the clinical trial program.

Novartis is actively expanding the program moving forward into the next phase of development of this agent in patients with CML resistant to interferon. An international multi-center phase II study is planned that will open for enrollment in January 2000, if not sooner.

Currently limited by availability of drug supply, Novartis has devoted substantial attention to making sufficient quantities of the agent available as soon as possible. These actions include moving production of STI 571 directly to facilities usually utilized for commercial scale manufacture and increasing the technical resources and capacities devoted to the product. Novartis strongly believes that these efforts should make available more than sufficient supply of this agent for the expanded clinical trial program.

Suzan typed the letter verbatim on her computer, and in the header to the e-mail wrote, "We got it! We did it!" She was excited and so were members of the support group.

Brian Druker called again to congratulate her on the letter. He gave her the news that he would have a fresh supply of Gleevec in December. He would start three trials at his home base in Portland, Oregon, one each in December, January, and February, and he told her that she was welcome to join any one of them.

≋

To this point, the public had not been officially informed of the remarkable Phase I patient trials related to Gleevec. That would soon change.

December 1999 marked the first time that the trial results were reported in a public gathering. The venue was the American Society of Hematology, known simply as "the ASH meeting, " in New Orleans, Louisiana.

The Holy Grail

Brian Druker was chosen to give the final talk at the plenary session, deemed by the organizers the most important speech of the day.

For weeks in advance, he prepared for the talk, reviewing every blood count and bone marrow report of each patient he would discuss, hiring a designer to create graphics on how the drug worked. He knew, well in advance of the meeting, that his would be a high visibility talk, he sensed that it would be the highlight of his career, and he wanted the talk to sound as polished and as clear as possible. He did not prepare a text. He knew the subject well enough to speak extemporaneously along with the help of the slides.

Over the Thanksgiving holiday weekend, Debra J. Resta of Novartis's clinical development staff reviewed the data one last time for Druker.

Druker and his staff crafted a very carefully worded press release on the occasion of his plenary talk. Concerned that a study coming out of Oregon would not carry nearly as much weight as if it had come from MD Anderson, Dana-Farber, or Sloan-Kettering, he made sure to include some quotes from prominent people in the cancer research field in the release. Accordingly, Rick Klausner, the head of the National Cancer Institute, was quoted, suggesting how important these findings were. We at Novartis were unable to make promotional claims about STI571 since the drug had not yet received FDA approval; so our comments did not appear in the press release.

Then it was time for his plenary talk. He was nervous, not at having to give the speech—he could handle that easily, he knew—but more at what the reception to the patient trial results would be. When he had spoken to an illustrious scientific crowd at Biarritz the previous summer, the patient trial results had received rave notices, so he was reasonably confident that the plenary session would not be dismissive. But he could not be sure. In truth, he had no idea how conference delegates would react. Though he considered his Gleevec work the "holy grail of cancer research," that did not mean anyone else would agree with him.

A few months earlier he had been genuinely concerned that the glowing words he would use to describe Gleevec would lead to a huge public demand for the drug; and that we at Novartis would not be able to meet that demand quickly. But as we had made clear in our response to Suzan McNamara, Novartis was giving very high priority to expanding our production capacity for Gleevec so that the patient trial program could be accelerated. We announced that the international multicenter Phase II study would begin enrollment in January 2000. We further announced that we would implement a centralized patient referral system.

So the Gleevec shortage issue was fading.

A Train Running at Full Speed

Druker's 20-minute presentation indeed proved the highlight of the conference—as well as the supreme moment of his career. His talk received significant media coverage, newspapers as well as the national networks. The media latched on to Druker's notion that the Gleevec Phase I trials represented a proof of the concept of molecular targeted cancer therapy. Among those interviewed was CML patient Judy Orem. According to Druker, ABC News had consulted its own leukemia cancer experts. They could have urged the network to tread cautiously; instead, buttressed by the reviews of their own experts, they expressed amazement at the Gleevec findings.

The public announcement of Gleevec at the ASH conference opened the floodgates. CML patients, for whom the Internet was not a prime source of information, could now read about Gleevec in their newspapers or learn about it from television coverage. It was now, in Brian Druker's phrase, "a train running at full speed." The drug supply issue was ending. The old concerns about who should go into patient trials, given the limited access and the limited supply of drugs, gave way to the new reality: Gleevec was becoming available in larger numbers and new Phase II trials were getting under way. Brian Druker alone had a waiting list of 200 patients. We at Novartis wanted to assure that as many people as possible received the drug. But initially our staff did not know what to say to these people.

In due course it became clear to us that we needed a better system for handling all of the phone calls coming in to Novartis seeking information about how to obtain the drug.

The turning point was the ASH conference in December 1999. Prior to that conference, Novartis Oncology's clinical trials phone line received 15 phone calls a month from patients and health care professionals. We knew that the announcement of Gleevec at ASH would heighten patient interest in the drug. There would be more Internet inquiries; our representatives in the field would be queried more. And the phone calls would come in to Novartis in larger numbers. We had to think about how to handle all this.

And indeed, during the ASH meeting, we were getting 2,000 phone calls a day. The call volume remained at that higher level after the conference for a month; then subsided to 600 a month.

Until ASH only three Novartis staffers were handling queries. We expanded our Call Center over the next two years gradually to 80 employees and gave fresh training to everyone.

≋

Patients and doctors sought us out to get their hands on the capsule. We soon realized that it would not be a problem to enroll large numbers in the planned Phase II patient trials.

Enrollment for the Phase II trials had actually begun in October

1999. The investigators were able to enroll 1,000 patients; to get that many normally would have taken up to five years. But with patients learning about Gleevec in the newspapers and on television, it took far less time in this case.

The goal of the Phase II trials was to get more safety data and to watch for any major side effects. The program began in countries that had not been involved previously: China, South Africa, Argentina, Canada, Singapore, Israel, and some European countries.

The inclusion criteria were the same as in previous trials.

The enrollment would carry through until the following June 2000.

One of the Phase II studies, incorporating 260 patients, sought to target patients not previously treated for advanced CML. Between 14 and 21 percent of the patients in the trials had a major cytogenetic response.

As part of another Phase II study for 94 CML patients in the blast crisis phase, 44 had received prior treatment for blast crisis and 50 were untreated. In the previously untreated patients, the overall response rates were 48 percent and 47 percent at 4 and 8 weeks of Gleevec therapy, respectively. In those who had received prior therapy for blast crisis, the response rates were 38 percent and 33 percent, respectively.

Overall, the Phase II results showed that the quick, very high response rate to Gleevec exhibited in the Phase I trials was reproducible; as the number of patients increased, the hematological response rate remained very high. The responses in the bone marrow rose as well.

Patients like Suzan McNamara have slowly learned how to exploit the great advantages of the Internet and a kind of patient elite has been forming, small at first, but getting larger and larger. These Internet activists are increasingly putting physicians on the spot for they (the patients) are sometimes better informed than their physicians. Patients become empowered and out of that empowerment, they begin to challenge their doctors on the best therapy for their diseases.

These newly empowered cancer patients are asserting a right to be healthy, no longer ashamed or embarrassed by their illnesses, insisting on gaining access to the latest and highest-quality information on their diseases.

Because patients feel a right to be healthy, they also believe that it is no longer acceptable to leave the field of medicine to the doctors. Patients want to learn as much as they can about their diseases; and to these patients there is no better way to do that than on the Internet.

For the physicians, the Internet seems sometimes less friend than foe. Brian Druker found out the hard way that his seemingly private comments to patients about Gleevec were repeated over the Internet almost immediately: "I'd say something to a patient and I would read about it on the Internet that night. I wasn't talking to a patient. I was making a public presentation." He became more cautious about what he told patients.

Powerful Tool

The Internet was certainly a powerful tool in the hands of patients.

When word began to emerge from the Gleevec Phase I trials about the efficacy of the pill, patients desperate to get their hands on the drug flocked to the Internet for information. In time, patients formed Web groups to compare their own experiences with Gleevec.

No other group of cancer patients benefited as much from the ubiquity and the immediacy of the Internet as did CML patients.

We had never seen so many patient requests for a drug, never so much demand. It was encouraging to us to find that there was such a need and desire for the drug. But there was a threatening side as well: If we were not responsive and did not succeed in distributing the drug quickly, these "Internet patients" could easily turn into aggressive activists reminiscent of AIDS patients from earlier years.

The problem for Novartis, as patients began to clamor for the drug, spurred on by the Internet, was the possible loss of control we would face over issues of efficacy and side effects if we simply turned the drug over to anyone who asked for it. We would jeopardize the whole development process that had, as its goal, obtaining speedy approval from the various authorities around the world. These authorities had high standards, one of which was that patient trials had to be conducted in such a way that it was possible to judge the efficacy of the drug.

We could not assure keeping to those high standards in the patient trials if we let everyone obtain the drug. Before we could provide broad access to the drug, we had to provide the scientific evidence that the drug worked and we had to be very precise as to which patients the drug worked on and which patients it did not work on. Our approach was to give anyone who asked for the drug the address where patient trials would be conducted and it would be up to the investigators in charge of those trials whether the patient qualified.

In saying no to patients who wanted the drug, we had to show great sensitivity. It's not just a case of saying yes or no. How do you say no, how do you demonstrate sensitivity? How do you assure the patient that you understand and empathize with his or her situation? Are you showing compassion? Of course, despite your compassion, you have to stay the course. We knew all too well of the experiences of other patients seeking other drugs. Those patients found no compassion. We wanted to avoid that situation.

Saying No to Patients

I had to say no to patients on various occasions and I wasn't happy doing it. In one case, the patient was a close acquaintance of mine. As much as I wanted to help this family out, I had to look the man in the eyes and tell him that we could simply not break the rules. He was a CML patient, but in the chronic phase and still benefiting from interferon. In that respect, he did not qualify for our trials. Nor did he qualify for compassionate use; his disease was not advanced enough. The man understood our dilemma and did not give me a hard time; still, I cringed at having to make such a decision.

The only way to show compassion was to answer directly and personally each and every letter or e-mail that came to me. We also made sure our own staff understood how volatile the situation could become if patients felt their wishes were being ignored. We told our employees, quite bluntly, that we had to create a dialogue with patients, we had to show them respect, and treat them as partners as they were fighting for their lives; or else we would have them on our backs. And instead of becoming potential allies, we would have enemies. We had enough examples from the pharmaceutical industry that showed how quickly and easily this could happen. Most of all we had to be honest. We had to explain to them what we could do, and what we could not do. In that way, we hoped we would gain their respect and understanding.

Ultimately, of course, we could not ignore the wishes of certain CML patients to obtain access to the drug even if those patients did not fit the criteria for patient trials.

Again, the basic criterion for entering those trials was either being resistant to interferon or being unable to tolerate its side effects.

We agreed that someone in the blast (final) phase of the disease, even if not wholly fitting the criterion for the trials, could get the drug on a "compassionate use" basis. For instance, the mother of a child with glioblastoma, a kind of brain tumor, got in touch with me to ask that we supply her son with Gleevec. It was truly a compassionate need. The child's disease was surely terminal as most patients die within a year of diagnosis. We arranged to get the child Gleevec. We had no patient trials operating at the time but could any of us as parents not empathize with this mother?

I have also tried to keep in touch with the patient groups on the Internet.

Norman Scherzer, the head of the Life Raft Group Web site which offers support for GIST patients, asked me to contribute a piece to the Life Raft Group Newsletter. It is Norman's wife Anita, a GIST patient, whom we profile later in the book.

In April 2001, I was happy to write a small article for the newsletter. Here is some of what I wrote:

There is real power in the Internet as a source of information, but especially as a way to connect with each other, to share experiences, knowledge and help each other. The Internet makes this all possible. These connections across countries and time zones with English as the common language have enlarged our horizons, and I would assume this has been a substantial help for all of you. Especially those who are suffering from a rare disease gain from the sense of community, and there is encouragement in the feeling that one is not alone with a given condition. Discovering that there are others, getting organized and helping each other with information and support is often critical, both physically and mentally.

We made substantial up-front investments in Gleevec. In addition, Gleevec is only effective today for a relatively small patient population. This means we will have to charge a relatively high price to recoup our costs and to allow us to continue to develop the drug, testing to determine if it would be effective in saving lives for patients with other cancers. Our price will be lower than the most expensive cancer therapies, but for many it will still be a high price.

As some of you may know, but for reasons which escape my understanding, Medicare does not reimburse oral cancer drugs. So there will be patients who cannot afford the therapy. At the same time it would be unethical to leave uninsured and indigent patients without access to Gleevec. So we will put in place a patient support program, hoping that through this measure people will not be denied therapy for financial reasons. At the same time we hope that, in view of the relatively small number of patients, insurers and governments will quickly agree to cover this innovative and life-saving therapy. On our side we will continue to investigate Gleevec's efficacy, looking next to possibilities in solid tumors. If we are successful in increasing the number of cancer indications and with it the usage among a broad range of cancer patients, we might be able to lower the price over a period of time.

≡

The story of Gleevec and the Internet goes back a few years. It begins with people like Ed Crandall, Peter Rowbotham, and Judy Orem. They were, so to speak, the pioneers and along with many others, helped to introduce Gleevec to the larger public.

The man who began to publicize Gleevec on the Internet was Ed Crandall. He was the second patient to enter Charles Sawyers's Phase I trials.

Ed had to start with a low dosage of Gleevec—50 milligrams—as part of the trials; unfortunately for him the dose was not high enough to control his CML. He remained stable for a while and asked to stay on the trial; but his white cell counts soon went above 100,000 and he entered the blast crisis phase and died that spring of 1999.

I must pause for a moment to dwell on Ed Crandall. It saddens me so much that, because we had not figured out the right dosage in time, we could not help patients in our trials sufficiently. The good news is that eventually we did come up with the right dosage to save many other lives.

Ed's death was a blow to the Gleevec program. Here was the very first public advocate of the drug, someone who had set up a Web site promoting the drug, someone with a visibility, who had not made it. By creating the first Web site devoted to Gleevec and putting reports from other clinical-trial patients on the site, he played an early and important role in the development of the drug.

It was just after Ed died that Peter Rowbotham began writing about Gleevec on the Internet.

Peter's wife was diagnosed with CML on December 12, 1997. That was the first time that Peter heard about CML.

That was the start of his quest for knowledge about CML—and later Gleevec.

It was in 1998 that Peter Rowbotham learned about Gleevec from Brian Druker's March 1996 article on the subject.

On June 1, 1999 Peter's wife entered the Phase I trials.

Peter's posts to Internet CML lists became one of the principal sources of information about STI571.

Throughout 1999 Peter Rowbotham was writing the major part of all the material on Gleevec that appeared on the Internet. Peter was writing for E-groups.com / CML / listserve and a man named Jerry Mayfield would take Peter's material and put it on his own site.

In time there would be 42,000 messages on E-groups.com / CML / listserve (later the Yahoo! group).

The great majority of Peter's research is conducted over the Internet, which provides access through Medline to the abstracts, and often the full texts, of more than 11 million articles in some 4,000 medical journals.

Patients become activists, Peter Rowbotham maintains, because if a patient thinks he or she is confronting a life and death situation, the patient will seize on any new developments. There is an enormous hunger for information about new drugs, therapies, cures, etc. because it is a life and death situation. Doctors tend to focus narrowly on their own area of interest. Patients, on the other hand, are open to anything.

≋

Patient activists have changed the whole doctor-patient relationship.

Let's take another look at one of those patients, Judy Orem, whom we met earlier in the book.

For three years she edited the *STI Gazette* for Gleevec patients at Oregon Health and Sciences University. She also started a study group in 1999 that met once a month.

Throughout the patient trials she became friendly with Peter Rowbotham. Judy also met Suzan McNamara, who sent the petition to us at Novartis. Judy became friendly with the other patients; after all, "we saw each other all the time." Friendships were formed. Judy became a kind of social chair for the group.

In the summer of 1999, it was time for a number of the study patients to go home. Judy and friends thought a farewell party was appropriate. Twenty-three people came to her house for lunch. The feeling of camaraderie was so strong that study people wanted the format to continue.

Informal Meetings

Starting in 1999, a group of Gleevec people began meeting informally for 90 minutes the first Monday of each month, the first two times at Judy's home and then in a room designated for them at Oregon Health and Sciences University. The Oregon Chapter of the Leukemia and Lymphoma Society supported the group by providing mailings, refreshments, and personnel at the meetings. The meetings attracted as many as 60 people, including patients, friends and relatives. The doctors and nurses leading the Phase I, II and III studies at OHSU came to visit and answer questions. Questions were put to the doctors, including Brian Druker, about the disease and treatment, side effects of Gleevec, etc.

Around the same time Judy began her newsletter, sponsored by the Leukemia and Lymphoma Society's Oregon Chapter.

In December 1999, the television show *20/20* interviewed the study group that Judy started, for a piece about Gleevec. The 16 study patients spent two hours chatting about the new drug study; the final TV version had them on for less then two minutes, but they had a wonderful two hours of laughing and crying together as they prepared for those two minutes.

As the year 2000 began, there had still been relatively little use of the Internet to spread the word about Gleevec. It was too early. Too few people had actually taken the drug. "We were afraid to say much on the Internet," Judy recalled, "because we were all doing well." Trial participants were reluctant to impose on others who had yet to take Gleevec the burden of hearing how well these Phase I patients were doing. "We didn't want those who were stuck on hydroxyurea or interferon to be upset with us."

So Judy and her fellow Phase I patients held back.

Judy Orem began to notice that on the Internet little nuances started to creep in between those who were on Gleevec and those who were not: "As the number of us on Gleevec began to grow, those who didn't get it were a little resentful." They felt the Gleevec patient trial members had taken over the support group site. Much of this had to do with the fact that there was still very little of Gleevec being produced.

When Judy was first put on the drug, she and the other trial members were told that they would take Gleevec for three months; then they would go back to something else to maintain their white cell counts while investigators worked out what the proper dosage was for patients to take.

But the investigators received such good results that there was no way they were going to take those first patients off the drug. And, as Judy noted, "all of a sudden you have a drug company (Novartis) that has pressure added to it to supply the drug. Novartis had to keep up with that kind of pressure."

Judy notes how hard it was to be on the drug knowing that there were many others who couldn't get the drug. "So we became advocates for these people."

Norman Scherzer, like Peter Rowbotham, like Suzan McNamara, like a number of others, has been a driving force in spreading the word about Gleevec to Internet enthusiasts.

The Life Raft Group that Norman helped to start (and has since become the leader of) has provided various people, patients, doctors, investigators, with a unique kind of data bank that cannot be replicated anywhere else, not even in patient trials.

As Norman Scherzer describes it, the Life Raft Group is a multi-faceted organization/community of GIST-Gleevec Clinical trial patients/loved ones. Life Raft Group members collect their own data and publish the findings in their monthly newsletter. "We've turned the clinical research world upside down," suggests Norman. "It can take years for patient trial data to filter down into the general medical community and even longer to reach the general public; the Life Raft Group offers that data in real-time. "When we look at side effects, however, we believe that we are more sensitive and accurate in reporting on this from the perspective of the patient and that we offer timely information to doctors as well as patients, particularly in the early stages of the clinical trials."

A major goal of the Life Raft Group, as Norman notes, is to help patients and doctors understand the nature and the management of the side effects of Gleevec. "Our ability to collect timely side effects information from a patient's point of view," he comments, "cannot be replicated by the medical community. We are the front line. We raise red flags."

Norman acknowledges that the most important reason for the Life Raft Group has been to help members get better. "It's a very self-ish reason," he says. "In the past one cancer patient would stand at the proverbial water cooler and ask another cancer patient, 'How are you doing?' We supplant that practice, but we provide many conversations and many water coolers."

The Life Raft Group has certain rules to protect its members. Its on-line discussion group, for one thing, is not a public site. To join, a member must fill out an application form. No doctors are allowed in. Patients want to be free to talk openly about a doctor without that doctor on-line. Patient confidentiality is jealously guarded when sharing information and patient identifying information is never included in any published data.

In an editorial in its June 2001 newsletter, Norman proposed that the clinical trial community establish a minimum effective dose of Gleevec instead of the traditional goal of escalating to a maximum tolerable dose. Norman wrote: "If Gleevec is truly to be considered a targeted drug, the natural extension of that concept is to adjust the dosage to the lowest amount required to bring the disease under control and the lowest amount required to sustain a lifetime maintenance level. We now know that when Gleevec works, it seems to do so very quickly, enough time to increase dosages on a selective case base where that seems indicated."

In its February 2001 newsletter, circulated to members and to clinical researchers at Novartis and at clinical sites throughout the world, Norman presented the first data showing the response of 16 GIST patient members to Gleevec. Response was defined as tumor shrinkage; he reported that 87.5 percent of the group responded to Gleevec, most very quickly.

In the June 2001 newsletter, he reported on the Gleevec response results of 38 Life Raft GIST patients on patient trials for at least two months. The data was based upon medical updates from 58 of the 60 evaluable member patients on the various Gleevec trials. Seventy-one percent of the group showed tumor shrinkage, with no differences among the three drug dosage groups—400 milligrams a day, 600, and 800.

Doctors occasionally contact Norman Scherzer to find out about new side effects of GIST patients. As an example, a number of the first 30 GIST patients who used Gleevec threw up, including Norman's wife Anita. The advice that Novartis had given was to take the pills with breakfast. Life Raft members fine-tuned that to spacing out the interval between taking each pill; eat a little, take one pill; eat a little more, take another pill, etc. The result was far fewer episodes of patients throwing up.

Norman Scherzer is adding new members to his Life Raft Group on a gradual basis. He wants the site to remain relatively small. "We have a psychology of community. Prior to us there was not a single disease group or organization related to GIST and certainly none related to GIST patients on Gleevec. We were unique patients and loved ones going through a unique experience and that takes a little time to figure things out."

According to Norman Scherzer, gender is not a good predictor of the disease and there are no gender differences thus far in how patients respond to Gleevec; however, females seem to have much more severe side effects than men.

A precedent for the efforts of the GIST Web site was the AIDS organizations that sprung up in the 1980s. AIDS galvanized the young, articulate gay community into action on a number of political and other fronts, with confrontation in the media and in the streets taking a lead role. This GIST-Gleevec Clinical trial community was, in comparison, miniscule in size and older and more mature in their ways. Even if all GIST patients were included (on Gleevec or not) the number would be only about 5,000 in the United States.

In early May 2002, Norman's GIST group held its first official gathering for all members at a Boston hotel. As I stood before the group to make some remarks, I was impressed with the courage and strength these people were exhibiting.

Managing Gleevec

6

Victory in May
A Record-Breaking Approval

We were all in a race against time. We felt that if we didn't get to the finish line quickly lives would be lost. Knowing the high stakes involved added to the tensions. We tried to keep calm. But it was not easy. After all, we knew that Gleevec was special.

The Phase I trials had first demonstrated that; then the Phase II trials validated those first trials.

Already, numerous patients were clamoring for the drug. The best that we could do was to put as many of them as possible into the clinical trials, but space was limited.

It was clear what we had to do. We had to produce the drug in large quantities. And we had to obtain the approval of the health authorities in the United States and elsewhere.

Only then could we truly meet the growing demand for the drug.

The man who took the lead in speeding up patient trials was a pediatric hematologist named Renaud Capdeville. He had been born in Paris in 1961 and later attended medical school there. He did an internship and residency in hematology and pediatrics in Lyon and later was the clinical leader for the development of the interferon program for CML at Hoffmann-La Roche in Basel. So CML was not new to him when he joined Novartis in 1997 as International Clinical Leader. He later took on a "group leader" position.

He became the STI571 clinical leader in February 2000. This was an exciting time. The public announcement of the Phase I trials at ASH the previous December had raised all sorts of positive expectations among patients and doctors. Those Phase I results, going far beyond the usual testing of how safe a drug is, had demonstrated that STI571 was already helping CML patients in their battle against the disease.

At the time that John Ford was handling the Phase I trials for Novartis, Renaud Capdeville was John's supervisor. Renaud would visit John's office down the hall and stare at the data on patients that John had put on his walls.

For Capdeville, it was only somewhat interesting when STI571 lowered patients' white counts, as it did in the early phases of the Phase I trials; but when patients began showing cytogenetic responses after only three months on the drug, that was indeed remarkable. He knew from his days at Hoffmann-La Roche that interferon sometimes resulted in a cytogenetic response in CML patients—but that usually took one to two years!

Normally Capdeville would have waited for the end of the Phase I trials before proceeding to the next step, Phase II trials. But when he saw that STI571 was leading to cytogenetic responses so quickly, he realized that it was important to move to Phase II much earlier than usual.

If, as was becoming apparent to him, the drug had a real potential to save patients' lives now, there was good reason to accelerate the trial process.

For that reason, Novartis began Phase II trials in June 1999, only a year after Phase I had begun in June 1998. On average in oncology, Phase I trials take two years; Phase II trials, one to two years—or three to four years for Phase I and II.

It was decided to begin two Phase II studies with patients in the most advanced state of the disease in a dozen sites in Europe and in the United States.

Very quickly after that some began to wonder why we did not study those patients who could not continue on interferon but are not in the advanced state of CML. This led to a third Phase II study in December 1999.

Expanded Access

The Gleevec trials were different in so many ways. There was the matter of the size of these trials, for instance. In 1999, with our decision to plunge ahead into Phase II trials came the question of how many people to put into these trials. Normally Phase II trials have only 50 to 100 patients, enough to allow investigators to learn a little more about the level of efficacy of the drug. But the CML patient community was coalescing. They were talking to one another—over the Internet mostly. Suzan McNamara had started her petition. The result was that a lot more than 50 patients were going to want to get into our Phase II trials.

We faced a growing demand from patients who wanted to get their hands on the drug. It became important for us to come up with ways to meet this patient demand quickly. That was why we did not want to close the Phase II enrollment before we had an arrangement in place that allowed patients access to the drug after the enrollment closed. Some patients were simply too far from the hospitals where trials were occurring and we wanted to accommodate them. We decided to consult patients and we did so at a meeting in Washington, D.C. in May 2000. We asked AIDS experts to attend and give us advice as well.

Our options were to create a "compassionate use" program in which we would give the drug to patients relatively freely without any protocol; or we could have done more studies and opened them to a larger number of patients. At that time we had some concerns about the safety of the drug, which dimmed our enthusiasm for the "compassionate use" approach.

In the end, in June 2000 we decided to launch an Expanded Access Program—which was something in the middle of these options. In a typical Phase II trial, the requirements dictated by the study protocol limit the number of patients who can be enrolled to a predefined number. In the case of the Expanded Access Program, the objective was to set up a flexible mechanism to allow access to the drug to a much larger number of patients than usually enrolled in a "classical" clinical trial, while ensuring an adequate level of safety monitoring. To achieve this, the program used a simplified version of the protocols of the preceding Phase II studies. However, to guarantee patients' safety, the implementation of the program was limited to centers with a high degree of expertise in treating patients with CML.

Some 7,000 patients were enrolled in the Expanded Access effort. As the drug became more available, Novartis expanded the program from six to 32 countries.

≋

One of those patients in the Expanded Access Program is Nigel Douch.

Nigel Douch talks about the eight months after November 2000 with great animation in his voice. He has lived through nightmarish times: just when he thought he might have turned a corner, he would experience another setback. Then he managed to get into an expanded clinical trial for Gleevec and life began to look better: "Here was hope where none existed before." Speaking in July 2001, he acknowledged that "even if my remission breaks down next month, I've had a good eight months. I've had a lot of fun." Had he not gone on Gleevec, he would have been forced to take interferon and undergo repeated chemotherapy. "I would have been very ill and would have been half the time in hospital. If Gleevec has given me nothing else, it's given me a good time."

In mid-October 2000, as his dosage of interferon was being steadily increased, Nigel learned that an expanded access program for Gleevec had been announced.

Nigel was a somewhat different CML patient than the usual ones who were given Gleevec. CML patients diagnosed with the disease while still in the chronic (first) phase generally have another five to seven years before their CML worsens and they reach the accelerated (second) phase.

In Nigel's case, by the time he was diagnosed with CML he was no longer in either the chronic or accelerated phase; he was in the blast crisis (third) phase. He did not have five to seven years before the disease became life threatening. Doctors thought at best he had no more than two years to live. Nonetheless, thanks to the three stints of chemotherapy he had reverted to the chronic phase for a second time and that was where he stood when he began taking Gleevec in November 2000.

At the time Nigel began taking Gleevec, his doctor warned him that possibly the most that he should expect from taking the pill was a short remission.

The prediction was that in early May when he took the six-month bone marrow test to check on the effects of Gleevec, his Philadelphia chromosome percentages would drop from 50 percent positive to 25 percent positive. That would have been the best to hope for.

In the first few days after starting on the drug, Nigel experienced serious cramps in his legs and his hands, "the sort of cramps that throw you across the room like an electric shock." He was given something to control the cramps that worked after a short while.

He has had no other side effect since then, other than some nausea that he learned to control by taking the pills after a meal rather than before—as he had been told!

Three months into taking the pill, in February 2001, Nigel took a blood test that showed good improvement in his white cell count.

On May 9, 2001, Nigel was given a bone marrow test. He had been on Gleevec for a week short of six months. The results were totally unexpected: Rather than the 25 percent level that had been predicted, he registered zero Philadelphia chromosomes in the cytogenetic test and zero in the FISH test.

As of mid-December 2001, Gleevec was still holding off Nigel's CML with quite tolerable side effects. In October he went back to work "except my tendency to fall asleep during long client meetings is a bit disconcerting." In recent months Gleevec became licensed in England and so Nigel began getting it through the normal prescription route.

By the fall of 2002, the health authorities in almost every significant country have approved Gleevec. It was unprecedented for so many people to have access to a cancer drug so early.

We were learning more about the safety of the drug in this program. Based on preliminary results, we had increased confidence that Gleevec was very safe indeed.

For Novartis it had been a massive challenge to cope with such a large number of patients in such a small amount of time. We had all along communicated to the staff that we placed a high level of priority on Gleevec and consequently we assigned significant resources to the project. At certain times, we assigned more resources to the Gleevec effort than to any other drug project at Novartis.

In the Phase III trials, we took patients who had been newly diagnosed with CML, and had not been treated with interferon.

In the Phase I trials, investigators had checked whether Gleevec worked on patients who had failed interferon. The big question in the Phase III study was whether Gleevec would work better if it was used before any other therapy was tried.

We randomly allocated patients to take either Gleevec or interferon. Enrollment closed at the end of January 2001. Some 177 hospitals in 17 countries participated. Prior to Gleevec, CML Phase III trials usually included only 500 to 700 patients. But we have enrolled some 1,106 patients in this ongoing study. It normally takes three to five years to enroll patients for a Phase III study. We took only seven months. This time investigators were looking much more at survival rates than at how well patients did on Gleevec in the early stages.

Data from the Phase III study showed that Gleevec was nearly three to four times more effective in achieving a cytogenic response in the first-line treatment for newly diagnosed CML patients than the combination of interferon and Ara-C. Moreover, Gleevec significantly delayed the time it took for the disease to progress to the more advanced stage of CML compared to the interferon/Ara-C treatment. The significance: The results indicated that the earlier Gleevec is used to treat CML, the better the response.

The results showed that patients had achieved major and complete cytogenetic responses of 83 percent and 68 percent compared with patients who were given interferon and Ara-C, who experienced major and complete cytogenetic responses of 20 percent and seven percent. The complete hematologic response rates were 96 percent for those who took Gleevec and 67 percent for those who took interferon and Ara-C.

≡

As the new trials began, the Ringaskiddy team that was busy manufacturing the drug was working to meet their impossible deadlines. They could never be sure that they would reach each deadline, but they were determined to give it their best shot. Their reputation was riding on how well they accomplished their mission, but more important to them, they knew that each day brought them closer to saving someone's life.

The Ringaskiddy team met the August 2000 deadline, going beyond the goal of 1,400 kilograms, producing 1,536 kilograms of Gleevec, largely to support the Phase III patient trials that had begun two months earlier.

How did they move so swiftly?

Not by hiring extra people. That was considered too time-consuming. They would have had to be trained, and there was simply no time for that. Instead, employees were given a lot more tasks to perform each day than they had been expected to do in the past.

Employees became personally involved in the success of the drug. Some actually contacted members of Gleevec support groups on the Internet. Employees also pressed their bosses to find ways to speed up production of the drug, coming up with ideas and possibilities on using the equipment more efficiently and more quickly.

Firing Up the Troops

Andreas Rummelt traveled to Ringaskiddy on a number of occasions to fire up the troops. His staff had already made sure that employees in Ringaskiddy heard all the positive news about Gleevec that had begun to appear in the press. Our associates proudly showed off the articles to family and friends, boasting that they were personally producing this drug.

By this time it had become abundantly clear that the quantities of the drug made at Ringaskiddy would be earmarked for the commercial market. Already there were indications that Gleevec might effectively treat other types of cancer, such as GIST.

To meet the increased demands for the drug substance, Ringaskiddy took yet another look at its production schedule. It transferred two early steps of the drug intermediate to a Novartis facility in the town of Grimsby in northeast England. Focusing on the remaining five steps, the team set up its equipment so that it could keep the five-step process running continuously. All five steps were up and running in parallel by the end of October 2000. During the launch campaign, it performed the steps one at a time. As more and more positive information came in about the drug, a new target date loomed: Gleevec was likely to hit the commercial market by June 2001. Employees spent their Christmas holidays at the plant and worked a lot of overtime.

≣

The increased quantities of Gleevec being produced in the summer of 2000 allowed Novartis to move forward with its expanded access program. Within weeks it had thousands of patients enrolled. The main criterion for inclusion, as before, was that the person had to be a CML patient—in any of the three phases of the disease—that was resistant to interferon.

The new program began just as the Phase II patient trials were shutting down. We set up the expanded access effort as a bridge between the Phase II trials and the arrival of Gleevec on the open market: We knew that it would likely take another year before Gleevec would be approved for sale in the United States, and another three years in other countries.

From our perspective, the expanded access program gave us a better opportunity to analyze the safety profile of the drug, in effect, to make sure there were no new major side effects to Gleevec.

By October 2000 the program had begun in China, South Africa, Canada, Israel, Argentina, Singapore, and some countries in Europe. The various patient trials had so many people enrolled in them that by the end of 2000 nearly half of our clinical organization was working on the registration file that was required for the submissions we would make to the various authorities.

In the weeks before the 2000 ASH meeting, as more and more data was announced, interest in the drug grew proportionally. Call volume at the Novartis Call Center began to increase. We were getting thousands of calls a day. Now we were in a better position to help callers: We could direct patients to ongoing trials; sometimes callers asked for information about diseases that the drug did not target. We directed those callers to other resources.

Countdown to Approval

Novartis was about to make medical history—maybe.

To distribute Gleevec widely and quickly in the United States, the pharmaceutical company needed the approval of the U.S. Food and Drug Administration.

Meanwhile, patients continued to contract the deadly CML disease; other patients were going from the chronic phase of CML to the more deadly accelerated and blast phases; in the worst cases, CML patients died of the disease.

These patients wanted to get their hands on Gleevec, betting their lives on getting the pills before it was too late.

Gleevec had been generating more excitement than any other new drug—and with good reason. It appeared to offer a hope to certain cancer patients—specifically those with CML and a few other forms of cancer—perhaps of saving their lives but certainly of prolonging them.

We faced quite a challenge in getting the FDA on our side, for it prided itself on holding to the highest standards. It would go to any length to determine whether a drug should be allowed on the American market. FDA inspectors, all of them experts in their fields, would fan out to production sites of a drug, checking on equipment, poring through documents filed by pharmaceutical companies, looking for any signs of noncompliance with regulations. If the FDA discovered any deficiencies, it simply decided to reject an application, no matter how many millions of dollars the pharmaceutical company had spent on the drug's development.

Phase II Results Enough?

The approval process was time-consuming, deliberately so.

Yet, we believed that we had a life-saving drug on our hands and therefore we had to convince the FDA to accelerate the approval process.

When in 1999 we realized that there was a potential to get STI571 approved very quickly by the health authorities, we began to consult with the FDA and the European authorities. We wanted to know whether our Phase II studies might potentially be sufficient to get an approval provided the level of efficacy would continue as it had been.

So we began consulting with the health authorities at a very early stage; we were fortunate to get their input on how best to construct the protocols for the Phase II studies.

The American health officials noted that if the response rates continued to hold up, a fast approval would pose no problem. That was an unusual acknowledgment on the FDA's part: Usually its officials would not make a determination on Phase II results but would insist on waiting for Phase III conclusions. We simply did not want to wait that long and we found empathy among the FDA staff. Only under exceptional circumstances would the FDA approve a drug on the basis of Phase II results—exceptional circumstances being that the drug has already been shown in the early Phase I and II trials to work effectively and save lives. We were counting on the remarkable early results of the Gleevec trials to win the day in Washington, D.C.

The process of seeking FDA approval required a great deal of work. Safety and effectiveness data for more than 1,000 patients enrolled in patient trials from June 1998 to April 2000 had to be compiled, analyzed, and reported. We had to put together all the work that had been done before STI571 entered clinical trials. We knew this would monopolize the time and energy of many people at Novartis, but nothing seemed more important than speeding up the approval process.

The FDA had six months to review and decide if STI571 was sufficiently safe and effective to permit its availability by prescription.

In July 2000 the FDA granted fast-track designation to Gleevec. U.S. regulatory authorities grant such designations only for drugs that treat serious or life-threatening conditions and have the potential to address unmet medical needs or improve upon existing therapy.

Fast-track status is something a pharmaceutical company can ask for and be granted by the FDA any time during development, even as early as the time of the first clinical trial. This accelerated status gives the company easier access to FDA officials for consultation; it also allows for several parts of the documentation to be submitted as they become available before the official filing in order to speed up review.

Getting fast-track status was a step in the right direction, but it did not guarantee that we would win quick approval. There was still much that the FDA had to do before giving us a thumbs up.

SPRING 2001

On-Site Visits and Heightened Expectations

And so in the early spring of 2001, as FDA inspectors visited the company's sites around the world, Novartis senior executives waited impatiently for the big moment. Auditing the sites of patient trials, the inspectors reviewed charts of about one in five patients to ensure that the data had been reported to them accurately.

For all of our confidence about the efficacy of the drug, an air of drama surrounded the approval process.

Could we meet all of the rigorous demands of the FDA?

Would our production facilities and quality control processes satisfy the FDA inspectors?

Was there enough data to support our contention that Gleevec's approval should be quick?

While we pondered these questions, the FDA remained highly confidential in its deliberations. There was nothing unusual in that. For much of the time that we were submitting materials, the agency offered no hint of whether it was inclined to give its approval for marketing Gleevec.

Going on the Fast Track

Armed with the evidence that we had about the drug, we had every reason to believe that FDA approval would eventually come.

Even though Gleevec was on the fast track, we could not be sure that approval would come swiftly.

Nor could we know that if and when it came, we would be ready to distribute it as quickly and efficiently as possible.

New Drug Application Filed

The "NDA"—New Drug Application—for Gleevec had been filed on February 27, 2001, seeking priority review. Novartis had submitted its application to the FDA seeking market authorization for Gleevec for the treatment of patients with CML in all three phases (chronic, accelerated, and blast crisis) after failure of interferon therapy.

The FDA grants priority reviews for products that might offer a significant improvement compared to products already on the market, in short, the products must be innovative.

Priority review will shorten the actual review time of an NDA from submission to first action, such as approval or approvable, from 10 to 12 months down to six months (or less as in the case of Gleevec).

A New Man Takes Over

Robert Miranda took over the Gleevec registration project in the United States for Novartis in early March 2001. He came at a critical point. Miranda found in his early talks with FDA officials a good deal of excitement and enthusiasm for the drug. He would never work as closely with FDA officials on a drug submission as this one.

For him the shining moment, the point at which he realized that Gleevec was indeed special, came when the phone rang one day that busy spring. The man on the other end of the phone had originally called the FDA to express gratitude for Gleevec and officials there passed the caller on to Miranda. It was the son of a CML patient calling just to say "thank you for developing this drug and saving my mother's life." That has been the only call Robert Miranda has received from a relative of a patient, and he has not forgotten it.

Meanwhile, Miranda continued his daily conversations with FDA project manager Ann Staton. The FDA assigned a larger team to the reviewing process than was normal, one more indication of how eager its officials were to review the marketing application in the shortest possible time. Usually one FDA official was assigned to study safety and efficacy; in the case of Gleevec, one person was assigned to study safety, the other to study efficacy.

It was gratifying to have the FDA work so diligently on our case.

Dr. Richard Pazdur, director of the Division of Oncology Drug Products, Office of Drug Evaluation I, Center for Drug Evaluation and Research at the FDA, and his staffers, had his division's doors open to us all the time. Issues that normally took weeks or months to resolve were dealt with in a single meeting. We asked questions about how to crunch the numbers; whether we would need paperwork on certain matters. The FDA was very constructive in its approach.

Novartis also sent applications for approval to the European Union, Switzerland and other countries. In the weeks before the submission, Jörg Reinhardt, Greg Burke, and his team worked day and night to get all the data together, including patient record forms. The acceleration of the process put tremendous pressure on everyone.

MARCH 26, 2001

The FDA Grants Priority Review Status

On March 26, the FDA decided to grant priority review of the drug.

The FDA filing application was based on results from some 1,230 patients in 32 centers in five countries; it was supported by data from three Phase II studies that provided both hematologic and cytogenetic response rates. That data showed a major cytogenetic response in patients with advanced stages of CML. Patients with chronic phase CML after failure with interferon therapy achieved an 88 percent hematologic response and 49 percent overall major cytogenetic response.

Once in a Decade Opportunity

The FDA's helpful attitude encouraged us to work that much harder. This was a once in a decade opportunity to communicate what our company was all about. The message was simple—ours is a pharmaceutical company that knows how to innovate, that has world-class professionals on its staff, and that cares about patients; its ultimate aim is to cure patients of their diseases.

The challenge for us, as the FDA approval moved steadily closer, was to make sure we were in a good position to communicate all of those messages.

Everything had to be ready to go from the moment we received FDA approval.

Our goal—a seemingly impossible one at the time—was to ship the drug within 48 hours of the approval.

We had to have a clearly thought-out patient assistance program in place. I had insisted that such a program be developed so that no one would have to forgo getting the drug on economic grounds. I suddenly became very nervous about whether the program would be in place in time. We had, until now, assumed that FDA approval would come by the end of June 2001 at the earliest, possibly not until the fall.

Now, in the early spring, it seemed possible that the approval could be just weeks away.

I traveled to our American headquarters in East Hanover, New Jersey and listened to my colleagues discuss the details about the patient access program; we made some changes to make it fair and generous, introducing a sliding-scale contribution. (We discuss this program in greater length in Chapter 7.)

LATE APRIL 2001

What's in a Name?

One seemingly minor issue, but a thorny one nevertheless, had to do with the name of the drug. The FDA had become more sensitive to the names of drugs in the past year, increasingly concerned that people would mistakenly take one drug, confusing it for some other. There was concern that a pharmacist might misread a doctor's prescription or mishear a request on the telephone.

We had submitted the name Glivec to the FDA and assumed that there would be no problem; that was the name we wanted to give the drug worldwide. But toward the end of April the FDA informed us that it would not accept the European spelling of the drug, although earlier in the process it had agreed to the name Glivec. It felt there was a potential for confusing Glivec with two drugs already on the American market, both for diabetes—Glynase and Glyset. The FDA thought Glynase and Glyset sounded too much like Glivec. They wanted the name changed—knowing full well that we were within weeks of getting the drug's approval.

We're Not That Smart!

We had to come up with another name. This was no small problem. Routinely, it takes the FDA two years to accept a name for a drug. Under normal circumstances two years would not be a problem since the proposal for the name is usually put forward to the FDA during the time of clinical trials—and it can easily take five years for the drug to win FDA approval.

With Gleevec in full fast forward, we did not have the luxury of that amount of time.

Needless to say, had we been forced to change the name of the drug from Glivec to a completely different one, we would have been forced into a lengthy delay, which would have been very hard for us to swallow—not to mention for the thousands of patients who were waiting for the drug. Beyond that, we had little incentive in changing the name of a drug that had already gained wide recognition.

We came up with what we thought was a rather simple, ingenious idea: we would simply replace the "i" in Glivec and make it an "ee"—

Gleevec. In that way, hopefully, the FDA would feel people would no longer be confused with Glynase or Glyset.

The FDA accepted our suggestion.

Some thought we had an ulterior motive in mind when we changed the name. Indeed, a reporter once asked whether we had done so in order that there would be an extra letter and it would be easier for us to create an "800" telephone number. My Novartis colleague David Epstein, the President of Novartis Oncology, correctly noted: "We're not that smart."

The only rub was that we would have to change the name everywhere that the original name appeared for American packaging. In one moment of panic, we were concerned that the name Glivec appeared on the capsule itself! That would have required throwing out every single capsule designated for the American market and creating an entirely new batch. For a full hour, until someone confirmed for us that the name Glivec did not appear on the capsule, we were calculating the monumental costs we might be incurring. In the end, the cost of changing the artwork on the package insert and the box ran to no more than a few hundred thousand dollars—tolerable considering that we were changing the name of the drug almost at the last minute.

≋

Once the name of the drug was settled, our thoughts then turned to how quickly we could get the drug to the market once FDA approval came through.

When it seemed clear that FDA approval was approaching, David Epstein turned to Andreas Rummelt, and asked: "If we get FDA approval as early as we expect, do you think we could significantly reduce the time between approval and getting the drug to market?"

Normally the time to market after FDA approval is 11 to 14 days.

"We can do what we can," Rummelt assured him, but he did not sound too optimistic. He knew that the FDA could hold up the process simply by requiring Novartis to go through the normal paperwork that could delay things by a few days.

Epstein stood firm: "Well, let's challenge the team and ask for 48 hours."

All that Andreas Rummelt could say was, "This is a big challenge."

Epstein's advice was to ignore the bottlenecks that the FDA might cause and simply assume that the FDA paperwork would not impede things very much.

Rummelt then turned to his own team and got a chilly response when he asked them to accelerate the post-approval phase of getting the drug to market.

The challenge seemed impossible.

"What if we forget about the FDA and its paperwork?"

Atef Adly, head of Pharmaceutical Operations U.S. and Jim Edwards, head of Supply Chain, conveyed the target to the U.S. Pharmaceutical Operations launch team.

The team began to soften: "It's tough but we'll accept the challenge. It might be possible if the FDA plays along with us."

MAY 1–4, 2001

A Final FDA Audit

Meanwhile, in Ringaskiddy, the pressure was building.

Further production was necessary right up to the final audit on May 1–4 to ensure that sufficient data was available to validate the manufacturing process.

We had our own team in place to deal with an FDA approval, should one come in the near future. Jim Elkin is the vice president for government relations at Novartis. He works in Washington, D.C., and his job is to keep abreast of governmental matters relating to Novartis. Gloria Stone is the public relations coordinator for the Novartis oncology business. She works in East Hanover, New Jersey. It would be these two people who were on the front lines for us.

MAY 7, 2001

Possible Decision in a Few Days

Gloria e-mailed Jim that she had been in touch with the National Cancer Institute and with the press office of the Department of Health and Human Services (HHS): Word was that there might be a press conference to announce FDA approval of Gleevec in the next few days.

It would have been highly unusual for the HHS to hold a press conference to announce FDA approval of a new drug. No one could recall the department ever holding a press conference for such an announcement. Normally, such approvals would be announced in a written press release.

HHS was keeping the news top secret, most likely because even its top staffers could not be certain when an announcement on the Gleevec approval could be made.

Picking up Rumors

Jim and Gloria rightly felt that we at Novartis needed to know as quickly as possible if the approval was truly imminent.

Jim put a phone call into Bob Wood, the Chief of Staff of DHHS Secretary Tommy Thompson. "I know you can't confirm what I've heard. I know it's very secretive. But I'm picking up these rumors about a potential press conference."

Jim Elkin pressed home one key point to Bob Wood: "If there is going to be a press conference, as the discoverer of the drug, we'd like to be considered for involvement in the event."

It just so happens, Jim told Wood, that Dr. Vasella *will* be in town on Thursday (May 10).

Remaining coy, Bob Wood would only say: "Well, I don't know that any such event is occurring. But, if such an event were to occur, I will go back to our staff and we will talk about what you're asking."

Gloria Stone meanwhile filled in David Epstein on the rumors, and Epstein in turn called me.

Meanwhile, on top of the Washington, D.C. rumors, my week could not have been busier.

In two days—on Wednesday evening, May 9—I was due to speak to the board of the Minneapolis-based medical device firm of Medtronic in Lausanne, Switzerland. Medtronic, the world leader for medical devices like pacemakers and heart valves, took its entire board to Switzerland where it has its European headquarters. The chairman of Medtronic was Bill George, a Novartis board member, and so I really wanted to give a good speech. I normally give a speech like this one a good deal of thought, and in the few days before the appearance, begin to write down my thoughts. But the press of business kept me from getting to my computer.

On that same Monday, May 7, we announced an important step for the company: purchasing 20 percent of Roche Holdings AG, another large pharmaceutical company based as we were in Basel, Switzerland. We had agreed to pay $2.78 billion.

Roche had once been the largest and most profitable drug company in the world but it recently had lost ground, due to sluggish sales in its pharmaceutical division.

I had received a phone call six days earlier, on May 1, from Martin Ebner, the second largest shareholder of Roche. He told me that he planned to sell his shares, representing a 20 percent stake in the company, and he had a buyer.

Did Novartis want to make a bid as well? he asked.

Chess Moves

We would not have been pleased to find that a competitor was about to purchase a large share of Roche.

While others guessed that we eventually wanted to establish a merger, I was content with getting a foot in the door.

It was like moving a figure on a chessboard, and by doing so being able to influence the game; within 24 hours we had made an offer and wound up buying the shares for $2.78 billion. We wound up with 20 percent of the voting rights. It was not only a smart move for us, but also an important signal that we wanted to avoid Roche being attacked by a rival.

On the day we announced the transaction, I tried to calm down any speculation while being truthful and fact-bound, saying: "This is a long-term financial investment, which is also strategic in nature. No collaboration has been discussed with Roche management, although we hope over time we will be able to explore areas of collaboration as Roche has good long-term business prospects."

May 7 was hardly over. At 6:30 p.m. New York time (12:30 a.m. Tuesday in Switzerland) I called Jim Elkin for an update on the possible timing of the FDA approval for Gleevec. He had just been talking to Bob Wood. Jim told me that we still did not know if there would be a press conference. And he explained that if there were to be a press conference, it was unlikely that Novartis would be represented.

Only representatives from the FDA and the National Cancer Institute were likely to be invited to be on stage at Secretary Thompson's side as he made the announcement.

This was not welcome news.

I felt such a format would not be fair as it excluded from the press conference the very people who had researched and made the drug—Novartis. I decided that, if that turned out to be the DHHS's final answer, Novartis would hold its own press conference.

Encouraging Jim to stay in touch with Secretary Thompson's people, I made it clear that I very much wanted Novartis to be part of the press conference. The fact that it would have been highly unusual to have the CEO of a pharmaceutical company present for a drug announcement was of no great consequence to me.

Pondering a Visit to the United States

All through May 9, Jim Elkin was trying to learn what he could from Bob Wood and some other DHHS staffers. But things remained fluid and nothing was confirmed. And before I could get on a plane to Washington, D.C., I had to deal with some other pressing matters.

First was what to do about the off-site executive committee meeting with spouses that we had scheduled for the weekend of May 12–13; I felt deep conflict about my participation in the event. If I decided to go to Washington, D.C., I seriously doubted I could be back for the executive committee meeting.

With it quite possible that Gleevec would be approved later in the week, I wanted very much to be in the United States. Yet, I felt a strong obligation to attend the executive committee meeting. My only choice, I supposed, was to postpone the executive committee meeting; I wanted to raise my dilemma with my Novartis colleagues.

I was pleased to find that they understood the conflict I felt; they encouraged me to head for Washington, D.C. We could always hold the executive committee meeting at a later date; but the FDA approval would be a unique event and they felt that I should be on hand.

The executive committee meeting was not my only conflict with the trip to the US. I was still scheduled to speak in Lausanne that Wednesday evening to the Medtronic board of directors. I would have liked to spend the entire evening visiting with Bill George and his board colleagues but I had to make a compromise.

I gave the speech and then chartered a plane to Washington, D.C. It was only late afternoon, May 9, New York time: Jim Elkin and Gloria Stone began to get word confirming that the press conference would take place the next day.

There seemed little doubt the FDA was ready to go forward. Its officials felt very good about this drug and were highly aware that patients were waiting. Proof of this was the FDA's willingness to take the highly unusual step of approving Gleevec within one week after the drug substance and dosage form inspections. The entire inspection and approval process took no more than two weeks. The usual approval process comes three, six, or nine months after such inspections. Speed certainly seemed to be of the essence.

MAY 10, 2001

A Historic Day for Novartis and Gleevec

2:30 a.m.

A Voice Message on the Cell Phone

As soon as I got off the plane in Washington, D.C., I turned on my cell phone to listen for messages. Jim Elkin had left one saying that the press conference was on for that morning and that I would be participating.

From what I could piece together, the Secretary had reacted positively to our being represented.

8:30 a.m.

Breakfast Preparations

The Novartis press conference "team" gathered at the Four Seasons Hotel in Georgetown, where I was staying, for breakfast and for some last-minute preparation. David Epstein was there, as was Burt Rosen, senior vice president for government relations and corporate communications; Kathy Bloomgarden, CEO of the public relations firm of Ruder Finn, Jim Elkin, and Gloria Stone.

All through breakfast, we edited the draft of the remarks that I would make at the press conference. I wanted to make sure that we focused on the breakthrough nature of Gleevec and the speed with which the company was bringing it to market. I also wanted to communicate the excitement and the dedication that Novartis employees felt toward the drug.

We discussed how to handle possible questions, including questions on the pricing of Gleevec; our theme would be that the cost of the drug was comparable to existing cancer therapies like interferon. We also wanted to be ready to talk in detail about the patient assistance program.

Meanwhile, Jim got on the phone to Bob Wood at DHHS to confirm that everything was in place for the press conference.

11:00 a.m.
An Historic Press Conference

The news conference itself was unusual; normally FDA drug approvals did not warrant such an event, especially one presided over by the Health and Human Services Secretary. But Secretary Tommy Thompson had decided to announce the marketing approval by FDA, recognizing that Gleevec was a breakthrough cancer medication that created new hope for cancer patients by opening new cancer research avenues.

Arriving at the Department of Health and Human Services at 200 Independence Avenue, we were escorted to Secretary Thompson's office. As I walked into the office, I thought of all the hard work that thousands of people at Novartis had done to get us to this moment. It was a truly exciting morning. The secretary greeted all of us warmly. He spoke about the breakthrough nature of the drug and about the upcoming announcement. I thanked him for including us in the press conference and we reminded each other that he had visited Novartis in Basel, Switzerland, in the past. At the time he was governor of Wisconsin. Ironically, at that earlier meeting, we actually discussed Gleevec in the context of our drug pipeline, but I had not mentioned any specifics as we did not yet know much about it. Little did I realize that he would become the Secretary of Health and Human Services and that he would one day (today!) announce FDA approval for Gleevec. I remember being struck in our original conversation in Basel with how vocal and engaged he had been on behalf of the American patient, academic research and scientific progress.

At the press conference, held in an auditorium at the DHHS building, David Epstein joined me on the stage along with Secretary Thomson, Dr. Richard Klausner, director of the National Cancer Institute, Dr. Bernard Schwetz, the acting commissioner of the FDA, and representing patients taking Gleevec, CML patient Suzanne Drager.

Seated in the first two rows were FDA personnel who had reviewed our file and finally gave the approval to Gleevec.

The FDA had behaved with great responsibility and great professionalism. We learned only at the press conference that many of the FDA reviewers had given up evenings, weekends, and soccer games with their children to do what they could to review the Gleevec file quickly.

Standing at the press conference I felt a rush of thoughts coming to mind. But the most powerful was of my sister Ursula. I thought back to her struggle and how deeply we had prayed during her illness for a drug that would help her Hodgkin's disease in the same way that Gleevec was helping CML patients.

Standing next to Secretary Thompson, in my role as Chairman and CEO of Novartis, I thought of the day that she died and again felt some of that same sadness and grief.

But, as the moment arrived when the cabinet secretary announced those words we were all waiting for—the FDA's approval of Gleevec—I thought of all the scientists who worked so diligently dating back to the 1960s to better understand CML. I especially thought of Alex Matter who by organizing his tyrosine kinase program in the early 1980s gave a kick-start to the effort that we were celebrating here in Washington, D.C. I thought of those patients such as Marco Nese, Judy Orem, Darlene Vaughan, Suzan McNamara, Sharon Godfrey, Nigel Douch, Nora Flanzbaum Friedenbach, who had benefited from Gleevec; and all the thousands of other cancer patients who would benefit from this drug and had helped directly and indirectly in the development.

There were plenty of kind words for Gleevec. Those who spoke shared our own enthusiasm about the drug and its potential.

Dr. Richard Klausner, director of the National Cancer Institute, declared: "This new drug, we believe, is a picture of the future of cancer treatment."

Secretary Thompson noted that the drug was based on the principle of molecular targeting, killing leukemia cells while leaving normal white cells alone. "We believe such targeting is the wave of the future," he said.

It was, according to Secretary Thompson, record approval for a cancer drug. "By reviewing Gleevec in just two and a half months, that is an all-time record for a cancer drug and for the evaluation of a highly complex novel drug."

In his remarks, the Secretary told of how he had lost three close family members to cancer. "So anytime you have a breakthrough, it is really a red-letter day."

Then, I was so delighted to hear him say—after all the concern, all the phone calls—"And I can't tell you how pleased I am to have the head of Novartis here who flew in, arrived at 2:00 this morning . . ."

Thompson showed wisdom in dispensing credit for the discovery of the drug. At first he noted that many people from many institutions contributed to the discovery. More specifically, he included both Novartis and Brian Druker for their roles, noting that it was "Ciba-Geigy in Switzerland that started us down this road and discovered the molecule that Dr. Druker from Oregon proceeded to find a way to put into a medicine."

Then I spoke:

"For me personally, this is an especially meaningful occasion. As a physician who has treated many patients, I know what a life-saving drug means for patients who are deathly ill and for their families.

"Speed was an absolute priority for Novartis in developing this drug. Upon the first hint of the dramatic potential of this new agent, Novartis rapidly invested extraordinary manpower to scale up manufacturing and to expedite the clinical development, allowing many,

many more patients to enter clinical studies and have access to the drug. We chose to take a significant risk at an early stage, reallocating resources and prioritizing the development of this product. As a result, the New Drug Application was filed only 32 months after the first dose in man, more than halving the typical drug development time of approximately six years. Based on the dedicated efforts of my Novartis colleagues and a constructive collaboration with FDA, who granted Gleevec a priority review, we have succeeded in bringing this revolutionary drug to the patients in record time.

"Patients' access to Gleevec has been one of our key concerns . . . and, today, we have about 8,000 patients around the world under treatment. We decided to put in place a comprehensive patient assistance program, which insures that uninsured, low-income CML patients are not denied therapy for economic reasons."

Representing the Patients

Suzanne Drager, from Falls Church, Virginia, represented the patients at the press conference. It seemed only fitting that someone who was actually taking Gleevec should be there with us at this big moment. She had been taking it since June 2000. She told of how she had been diagnosed with leukemia in April 1997 and sought a bone marrow transplant, but had no compatible donor. She went on interferon and Ara-C, but after two and a half years the drugs no longer kept her leukemia stable. Through her doctors she sought to get into a patient trial in Oregon under Brian Druker. Within three months she had gone from 100 percent Philadelphia chromosome positive to 21 percent. Before taking Gleevec, she had a hard time getting out of bed some days. Since taking the pill she had gone back to work full-time, "just basically picking up my life where I had left off four years ago when I was diagnosed. It's been a great year for me. It's been wonderful. I feel great." She has had relatively few side effects; just "minor annoyances. I'm headed toward a full remission. I'm that far away from it."

With the news conference and with all of the major news organizations running stories about the FDA approval, it was no wonder that May 10, 2001 seemed very special.

It was Gleevec's debut, the day the pill was truly born.

A little less than three years had passed from the start of the first patient trials in June 1998 to the FDA's approval.

≋

We at Novartis were elated. Now we had the strongest possible evidence that all of our hard work in accelerating the production of Gleevec had borne fruit.

We would be able to get Gleevec into patients' hands immediately.

≋

Rightfully, Secretary Thompson and I thanked Dr. Pazdur and his FDA team for the enormous effort they had put into the fast review of the Gleevec file. David Epstein remained behind to field more press questions and meanwhile I was escorted back to the secretary's office for a kind of debriefing on the event. The secretary was especially interested in what Gleevec might mean for other cancer treatments and I filled him in on a number of the patient trials we were conducting to investigate those very questions.

It was around that time that I had a chance to read the FDA's news release on the day's events. This was indeed an historic document, announcing "the approval of Gleevec (imatinib mesylate, also known as STI571), a promising new oral treatment for patients with chronic myeloid leukemia (CML)—a rare, life-threatening form of cancer."

According to the release, Gleevec has been shown to substantially reduce the level of cancerous cells in the bone marrow and blood of treated patients. The FDA noted that the trials had not been designed to determine whether Gleevec improves survival. The side effects reported frequently in the trials included nausea, vomiting, edema (fluid retention), muscle cramps, skin rash, diarrhea, heartburn, and headache. Severe fluid retention occurred in up to 2 percent of patients.

It was only that evening, back in the hotel that the full impact of the last few hours dawned on me. I turned on the television to C-Span and the entire press conference was being shown! I was really pleased at the amount of visibility Novartis and Gleevec received on this day.

But I remember feeling that, in all the excitement, I did not want any of us to forget that we were still in the very early stages of testing the drug on people. It was of course gratifying that suddenly, the world was beginning to learn about Gleevec. All the major news organizations ran stories after the news conference. In many ways, this day had been like the blossoming of a flower; but I reminded myself that we had to remain cautious and make sure to dampen any false hope that Gleevec would be a wonder cure for any kind of cancer.

Time to Celebrate

For Novartis, it was time to celebrate—from the people in the laboratories to the production workers.

Jörg Reinhardt was addressing 200 investors at a conference organized by Exane, a French investor group, in Paris on that day. He knew that the FDA's approval would come at around 5:00 or 6:00 p.m. French time, so that afternoon when he took questions after his talk about whether he had any news on the Gleevec approval, he had to act innocent. There were no celebrations that evening in Basel. But we would sponsor various company celebrations in the coming months. May 10, 2001 was a special day for Elisabeth Buchdunger, the researcher who had played such an important role in Gleevec: "There were many months of hard work, and all the years before as well. We were told all the time that our workload would improve at some point but it never ends. So, we should learn to celebrate some things and the FDA approval was one of them, and in record time. We've got approval in a number of countries in the meantime. It's exciting, but the first one, the big one that was the real excitement."

May 10 was no less special for the employees in Ringaskiddy. Having gone through that very stringent FDA audit in early May, the employees felt a strong sense of accomplishment. There was plenty to celebrate. Plant manager Jerh Collins joined a group of employees at a local bar for drinks: "It was fantastic, people were laughing, they were proud, talking about Gleevec all the time; there was a very, very strong feeling of value, a feeling that I made a difference. People held their chests out, that they worked at Novartis on Gleevec."

By the end of July 2001 Ringaskiddy had produced 16,500 kilograms of Gleevec. With FDA approval, Novartis staff labored to get the drug to pharmacies as quickly as possible, hopefully. within 48 hours.

MAY 11, 2001

Gleevec Starts Shipping

On May 11 at 9:00 a.m. the Novartis Drug and Regulatory Affairs FDA liaison Sharon Olmstead delivered the Drug Listing package, required of all drugs from overseas, to the FDA in Washington, D.C.

A half hour later the first batch of packaged Gleevec was shipped from the New York production facility to Novartis's East Hanover headquarters for distribution.

An hour later, the FDA approved and completed the Drug Listing, clearing the product for sale. (This process usually took seven to ten business days.)

At noon on that day, after receiving the Drug Listing notice, the Quality Assurance staff issued the Certificate of Analysis for the drug product and the Quality Compliance staff released the finished product.

Five and a half hours later, our Distribution staff shipped some 5,500 bottles of Gleevec to wholesalers, fulfilling all the on-hand orders.

A total of 40,487 bottles were packed and shipped to East Hanover for later distribution.

We had delivered Gleevec in a day! This was an unheard of feat in the pharmaceutical industry. That meant that by May 14 and 15 Gleevec would be available to retail pharmacies.

We gave bonuses to many people at Novartis for their work on Gleevec. It was one of only a half dozen times or so that we gave such bonuses for work on a selective project. The development and production of Gleevec was such a special case and we wanted to reward those who had contributed to the quality of the process and to its speed.

≋

By the day of the FDA approval, Novartis had submitted filing applications for Gleevec to health authorities in the European Union, Switzerland, Canada, Australia and Japan.

With FDA approval and with other nations' approvals expected in the near future, we were pleased that this tiny orange capsule was getting to all of the patients who needed the drug.

We never lost sight of the demand that patients were making for the drug. It was a clamor that kept us on our toes and seemed to build with each day.

7

Success Management
Giving Credit Where Credit Is Due

From my experience in overseeing the development of Gleevec, I have come to appreciate that there are even more complex and thorny issues that have to be dealt with when a product is successful than there are when it fails. Indeed, "success management," as I call this process, is challenging. We have a tendency to believe that, because a product has been approved by regulatory authorities successfully, the process of getting that product to market will automatically go smoothly and encounter no issues on its way.

And so, in this chapter, I want to address some of the challenges and pressures that Novartis faced as we "managed" this great success story—the discovery and development of Gleevec.

In fact, due to its uniqueness, Gleevec offers perhaps the best example through which to show the various challenges and pressures on Novartis and on the entire pharmaceutical industry in the quest for making better and better drugs.

Let's start with the greatest challenge of them all: the grudging feeling that many have toward our industry's financial success.

Few other industries need to justify the profits that they make from selling their products as much as the pharmaceutical industry. Many view health and access to health care as a human right. So it seems unacceptable that people suffer and die without getting the best treatment. At the same time the costs for health care are increasing as there are more and more senior citizens, consuming more and more medicines and health care services.

Even if hospitals and medical services make up a much larger portion of the health care costs, drugs and drug companies still come in for criticism. The industry, which in general makes healthy profits, is viewed as greedy. While people like our innovative drugs, we are sometimes accused of making a profit on the back of the patient. Watching their health care budgets too, governments exert pressure by setting up additional approval hurdles and imposing price controls and price cuts for drugs in many countries.

In one way we are just businesses, no different from a Microsoft, General Electric, or PepsiCo. We sell products and want to make a profit. Without profits there would be no capital, no research, no investments, and eventually no new products. Many people seem to ignore this fact, clutching to short-term focus; they just want to get cheaper drugs.

Part of success management is therefore to repeatedly explain the need for continuous investing in research and development, the need for patents, the need for profits, if we want a successful and sustainable pharmaceutical industry. It is vital to understand that there are no substitutes for the pharmaceutical industry. No government and no other institution has been as successful in discovering, in developing innovative drugs, and bringing them to the patients in need.

The cost of that research and development—now put at $880 million per drug—is steep. Marketing our products has also become more and more expensive. So the industry is caught in a bind between pricing pressures and higher internal costs.

Accusations

When we succeed in producing a breakthrough drug like Gleevec, we are immediately confronted with the danger of being accused that we will take advantage of patients who have little therapeutic alternative. We know this from the start.

All of the challenges to our industry have come together uniquely in the story of Gleevec: It is a life-saving drug, which created enormous patient and physician demand for the capsules, even before we had regulatory approval and while Gleevec was still in the early stages of development. It has received widespread publicity, swelling the demand for the drug. Because of the upfront R&D and production costs, the drug has had to be priced high, as there are only few patients who get CML and who require Gleevec. If we want to recoup our investment and make a profit, we have to price Gleevec at a level similar to the existing therapy, interferon.

The drug came along at a time when governments, patients and nongovernment organizations were complaining about ever-increasing health care costs. So there would be a lot of scrutiny, even if Gleevec would reduce the overall treatment costs by shortening the length of the hospital stay and lowering the number of doctor visits.

For all the above reasons, we understood that we had to treat everyone who wanted the drug respectfully. In my mind, no one should have to forgo Gleevec therapy because he or she could not afford it. So we developed a novel patient assistance program, which allows financially weak patients to get the product.

Our most strenuous critics leave the impression that they would like drugs to be distributed free of charge, or at cost; and they would like patents to be abolished so anybody could produce and sell cheap copies.

But an important part of our economic survival is tied to our ability to protect our patents. If we produce great breakthrough drugs like Gleevec, but we cannot enjoy the patent protection that allows us to recoup our investments and make a profit, the industry has no economic incentive to produce such drugs in the future.

Not helping our case to protect patents has been the international backlash against the pharmaceutical industry for its legal action against the South African government, which wanted to abolish patent protection for drugs, and later focused these efforts on AIDS medications.

These moves by the pharmaceutical companies to protect themselves have met with incomprehension. But what is generally overlooked is that patent law and potential profits are critical to the contemporary business model required for a productive pharmaceutical industry.

We at Novartis know we have been fair when determining our drug prices. But critics make it harder for society to believe that we are sincere in what we say—and in what we do. We hope that our behavior with regard to Gleevec, taking a huge investment risk despite a small patient population; accelerating our manufacturing process; creating a patient assistance program for this drug, will demonstrate that we truly care about patients. We can do so as long as we are successful, satisfying the interests of the shareholders by providing a return on investment. We would hope that once the general public learns that we are not simply heartless, faceless, and greedy industrialists, but people who care and want to help the patient population, then perhaps the same public will have a greater tolerance for drug prices and a willingness to listen to our arguments of why we must be allowed to make a profit.

The responsibility to produce drugs rests not with governments, not with academic institutions, but with the pharmaceutical industry. There is good reason for that. And we feel perfectly comfortable arguing that along with the responsibility for researching and producing those drugs must come a financial benefit as well.

No Accident

It is no accident that private enterprise has taken the lead in the drug-producing business. For it is only the pharmaceutical industry that has the incentive (admittedly, the incentive being profits) to spend the huge sums required to produce a drug.

Why has it been left to private enterprise to take on research and development and production responsibilities for drug making? Or to ask the question differently, why have academic institutions—where, after all, the brainpower exists in large quantities—not taken on those responsibilities?

The answer is clear:

If one looks at how many of these academic institutions are capable of bringing a product to the market, the answer is dismal. It is absolutely dismal. Why is that? For one thing, a very long time is required to produce a drug. In earlier years it took at least 15 years of development time and discovery; now it's perhaps 12. For another, academic institutions would have to devote time and huge financing to supervise every aspect of drug development. Of course researchers in academia do make important discoveries. But there is so much more to producing a drug than the pure research: all the pre-clinical work including the toxicity studies, the animal trials for efficacy, the testing for proper absorption, other issues dealing with distribution, metabolism, and excretion, as well as the side effects of a drug; finally, there are the patient trials. All of this requires a great deal of time—years in fact—and a great deal of financing—hundreds of millions of dollars.

Accordingly, one needs not only a knowledge base, but also an organizational structure, and a systematic approach. Certainly, perseverance is also needed. I do not know of any academic institutions that are set up for such investment and effort. The academic world is far better suited to the discovery process than to developing discoveries.

Only the pharmaceutical industry can take on the development of drugs in this day and age because the costs of discovering and developing products have been steadily increasing.

All around the industry there are pressures that force costs to soar.

There are the increasingly stringent and complex demands placed on pharmaceutical companies from the regulatory authorities. Patient trials include thousands of patients these days, no longer hundreds. More chronic diseases are being investigated and that means longer observation times in patient trials. Pushing the frontiers of medicine—coming up with new, more useful drugs becomes more and more difficult. Then there are new technologies, which have to be financed—technologies like combinatorial chemistry, proteomics, and genomics. Add to this the fierce competition in the market place.

We are all distressed at the high cost of health care, including the costs of acquisition of drugs; and we must make greater efforts to contain those costs. But it is folly to assume that pharmaceutical companies can continue to produce breakthrough medicine without someone financing these efforts and an illusion to believe that with the higher consumption of drugs by senior citizens, drug costs will not increase.

And so when it came to Gleevec, we wanted to have a chance to recoup our investment and to receive an adequate return on our investment; at the same time we recognized that Gleevec was unique in that it was unquestionably a life-saving drug. We therefore wanted to make sure that any CML or GIST patient could get the capsules regardless of the ability to pay.

It is against this background that we decided on the price of Gleevec. We knew that we had an uphill battle to win understanding for our decision. Striking the right price for this breakthrough drug has been one of the hardest parts of "success management."

In creating the proper price structure for drugs, we are guided by a set of objectives that we call our Purpose and Aspirations:

> We want to discover, develop and successfully market innovative products to cure diseases, to ease suffering and to enhance the quality of life. We also want to provide a shareholder return that reflects outstanding performance and to adequately reward those who invest ideas and work in our company.

At our shareholders meeting in 2001, I made the following comments on the pricing issue:

"... The focus is primarily on patients. We serve them by providing innovative, superior medications, which are effective against diseases that could not previously be treated adequately or at all. In order to keep on achieving this goal, we need the best, most expert and committed associates. At every level, they should be compensated according to their performance, in line with market rates. But you too, Ladies and Gentlemen, as investors, as the owners of Novartis, are entitled to receive a profit corresponding to the company's performance, part of which we plow back into the company and part of which we distribute directly to you.

"It's almost a platitude to say that, in order to achieve these goals, we need to invest consistently in R&D and in marketing. Taking all the various factors into account, a new drug will swallow up about 1 billion Swiss francs or $600 million [by December 2001 the figure had grown to $880 million] before it is launched. On average, about 10–14 years will elapse before it is ready to be marketed. As patent protection is limited to 20 years, we then have about 6–10 years to make a profit that justifies the risk and capital outlay and enables us to make the new investments that are required. In the absence of patent protection, a price cannot be charged which will yield a profit, and with no prospect of a profit there will be no investment in R&D. This has nothing to do with morality, good or bad; it's a matter of simple logic. In other words, in our market-based capitalist system, it is only possible to make long-term, high-risk investments—for example, in research—if there is a realistic chance of making a profit."

≡

That being said, there *are* demands put upon pharmaceutical companies to keep drug prices low. Try as we might, it's simply impossible from our perspective to keep drug prices as low as everyone would like.

As competition has increased among pharmaceutical enterprises, the cost of the marketing and selling of drugs has risen concomitantly; companies feel that they must deploy much larger and larger sales forces to be competitive; then, too, the new phenomenon of "direct to consumer" advertising, especially television, but radio and the Internet as well, has made for higher costs.

With new realities—people living longer; an inevitable rise in health care, as one gets older; the more effective treatment of certain "old age" diseases—comes a significant increase in the patient's consumption of medical treatment and services.

And that is an important reason why the launch of a product costs several hundred million dollars.

A Word About Medicare

In the United States, Medicare does not cover oral medications or oral cancer therapies. While most countries outside the U.S. have health care systems that pay for medications, the U.S. relies upon private insurance, Medicare and Medicaid.

What makes the American market so different—and so competitive—is the fact that drugs can be priced freely: Pricing is not regulated. Most countries operate like planned economies, where the government sets prices. But the United States does not. As a result, the American pharmaceutical industry has grown faster and is more profitable than in any other country. As a consequence, the American pharmaceutical industry has been gaining market share in its competition with European pharmaceutical firms.

≋

All sorts of factors have to be taken into account in setting up a pricing structure for a drug.

One is the potential market size for the drug. Others are the medical need, alternative treatments, and competition, just to name the most important.

It was certainly good news for Novartis that, in Gleevec, we had a drug that could not be substituted; for which there was no alternative; and that was potentially life-saving. In short, there was a strong market pull for Gleevec.

However, Gleevec targeted a small patient population. Each year, some 5,000 Americans over the age of 10 are diagnosed with CML. Worldwide, the figure is 1.3 per 100,000 a year. But the number of cases of prostate cancer and breast cancer is thirty times the CML figure.

Add to the mix another critical factor: Novartis had required significant up-front investment. The nature of the high risk that we took was not the hundreds of millions of dollars that we spent on the drug. We spend amounts of that size on any number of drugs. The high risk came from our decision to "front-end" a good deal of the costs, that is, to spend a sizeable portion of the costs of the drug in the early phases of the development process. Normally, we would spread the costs over the entire development and production process, taking a hard look at the drug's prospects for making it to the market and doing well each step of the way.

Given all these pros and cons, our calculation was that Gleevec would have to be an expensive therapy if we wanted to have a slight chance of making a payback.

But we were especially conscious of the cost of other similar treatments for CML patients. Interferon, the current standard of care for the 80 percent of CML patients who are not candidates for bone marrow transplant, costs $1,700–$3,300 per month in the United States. Elsewhere, the package insert dose of 9 million units daily of interferon therapy ranges from a low of $1,250 per month in Australia to $4,750 monthly in Japan.

Clinical trials have demonstrated that Gleevec offers patients greater benefit than interferon; patients in our studies represented those who responded to Gleevec but who no longer responded to interferon or could not tolerate interferon and had to discontinue therapy. That, in and of itself, gave Gleevec the advantage. In general, in chronic phase patients, Gleevec's major cytogenetic response rates are at least 65 percent. Although a range of response values has been reported in the medical literature for interferon, these generally have been lower than those in patients treated with Gleevec. Also, based on our review of the published literature for interferon, the approximate 90 percent hematologic response rate with Gleevec compares to a response rate of between 46 percent and 80 percent for interferon in the chronic phase of CML.

We also wanted to take into account what additional value Gleevec would offer. Certainly the greatest value was that it appeared to have a positive effect on CML in a relatively short amount of time. Beyond that, it produced tolerable side effects. Finally, it was taken orally in capsule form, avoiding the daily injections of interferon.

The Cost Structure

Weighing all of the various factors, we eventually created the following cost structure for Gleevec:

For treatment of CML in the chronic phase, at a dose of 400 milligrams daily, the cost will be $2,000–$2,400 monthly. For treatment of CML in the accelerated, or blast phase, at a dose of 600 milligrams daily, the drug will cost $3,500 monthly.

At the high end of CML treatment is an allogeneic bone marrow transplant surgery at $193,000 per procedure and a high complication rate.

A last resort in treating CML patients is chemotherapeutics— they fall into the same price range as interferon and Gleevec. The price of Gleevec is also competitive with other chemotherapy agents for other types of cancer, particularly when response rates and duration of response are taken into account. For example, combination therapy with Herceptin and Taxol for patients with advanced breast cancer can cost $37,500 annually, and the treatment is only 50 percent effective. The annual cost of combination therapy with the newer agent, irinotecan, and the older drugs, 5-FU and leucovorin for advanced colorectal cancer, is $28,125 but the therapy is only effective in up to 40 percent of cases.

Helping our case was the fact that CML patients would need fewer consultations with physicians and fewer hospitalizations, so costs would be saved through the entire health care system even if the drug was more expensive than some might have wished.

Should the price be the same for the drug in every country?

We're in something of a no-win situation here. No matter what we do, we open ourselves up to criticism.

The poorer countries argue against a single worldwide price, suggesting that it is unfair to have the same price everywhere because American purchasing power is greater than in Italy or in Brazil or in the African countries.

Obviously some countries wanted the price of Gleevec to be cheaper than $2,000–$2,400 monthly. But, in our view, that only encourages countries where the drug price is more expensive to import the drug from a country where it is cheaper. In these cases most of the price difference is kept as an additional profit by the distributing company, whether the wholesaler or pharmacist.

Furthermore, if the drug costs more, say, in the United States, than in other countries, Americans feel, justifiably, that they are underwriting the cost of the drug for other countries where Gleevec is sold more cheaply. Nonetheless, we are sometimes asked to adjust a drug's price to varying markets to assure that everyone can afford the drug.

In the end, we decided to create one universal price for Gleevec—and to put in place patient assistance programs to make sure that indigent patients could afford to get the drug.

The worldwide price was set at $2,200 a month for Gleevec.

It was clear that for some patients that price would create an insurmountable barrier to getting the drug. For the first time ever, Novartis organized a patient assistance program in the United States and in Latin America.

The company selected the following criteria:

Anyone earning under $43,000 a year received the drug for free.

Anyone earning between $43,000 and $100,000 a year would pay no more than 20 percent of their income for the drug. Hence, someone earning $50,000 would pay no more than $10,000 a year rather than the actual cost of $26,400 yearly.

Those who earned above $100,000 a year would pay the full price for the drug.

In the three salary categories parents with children would receive exemptions amounting to $5,000 per child a year. So, someone with two children would only pay 20 percent of his or her income for the drug if he or she earned between $43,000 and $100,000 a year.

One condition: The drug is not given free to someone who falls below $43,000 a year but has net assets that exceed $250,000 beyond certain exemptions such as a house and car. In short, someone cannot have other financial options and get the drug for free.

We have set these prices knowing that we will still get criticized. With so many different populations to please and so many arguments to be made, it is impossible to please everyone. In the end, we cannot become the servant of all masters and we have to make decisions.

A High, but Fair Price

We agree with those who say that the price we have set for Gleevec is high. But given all the factors, we believe it is a fair price.

While the price may seem high in absolute terms it will still work out to be less expensive than today's standard therapy, which in some cases involves prolonged expensive hospitalization. Then, too, in return for that price, the person receives a considerably enhanced quality of life and the life-prolonging effects of Gleevec. The price that we set enabled us to set up a special program to facilitate access to the drug for uninsured and needy CML patients. Finally, the price enables us to invest further resources in researching new cancer indications.

Although we have created programs to help people who cannot afford the drug, we cannot replace insurance systems nor should we be expected to pick up the slack created by the failure of governments to take care of their own citizens. We can assist and help in exceptional cases. And that is what we are doing. In creating the Patient Assistance program, we might incur a bit of heat from our colleagues in the pharmaceutical industry. We may be setting a precedent that they find uncomfortable. But if a drug is unique and life-saving and for a small patient population, there's every reason in the world to make sure that the drug reaches everyone who needs it."

We felt we had made the right pricing decisions. But we still needed to know whether others agreed that our decisions were correct. I wrote a piece in the April 2001 newsletter of the Life Raft Group noting that Gleevec was expensive; that it was a complex product; that we had invested large sums of money for a drug that at best had a small patient base. In the end the patients seemed to find what we had decided acceptable. In the United States we met with patient groups, walking them through the whole program; they were very supportive of our pricing arrangements.

Along with the pricing of a breakthrough drug, success management has to do with transmitting the story to the media carefully and effectively.

Transmitting the story of a breakthrough drug should not be too difficult. Or so it would seem. Scientists have achieved a medical breakthrough. They have developed a new approach to cancer therapy. In the laboratory they have created a drug that has the capability of turning the lives of patients with certain kinds of cancer around.

We are proud of our company and its staff, proud of what they have accomplished, thrilled that their work will bolster the image of the company.

So many people will feel good about the new drug:

The patients—some whose lives will be saved; others whose lives will be prolonged;

The physicians who now have some hope to offer patients;

Other scientists who can take the new approach used in Gleevec and try to make strides in fighting other cancers; and

The shareholders will be pleased that the reputation and revenues of Novartis should climb, as news of the new drug spreads.

Yes, there is plenty of reason to be happy.

It all seems fairly straightforward. But, alas, it is not.

Segmenting Audiences

We do not have the luxury of communicating exclusively with one audience. The message that we send to one group of listeners will inevitably reach a second group, and so on.

What we tell the financial community about the drug, the media will learn about. What we tell the media, will reach financial analysts, shareholders, patients and doctors. What we say to the financial community, will come to the attention of employees, and so on. What we say to shareholders, whether at the annual meeting or in the annual report, inevitably will reach the general public.

So there is no point in trying to communicate one message to one audience and a different message to another. It will not work.

As the management at Novartis searched for the best way to communicate the remarkable news about Gleevec, we felt we had to clarify exactly what the company stood for, what our own deep beliefs were. If we did not get these things straight first, we could easily get confused over what messages we sent out to different audiences. Very soon, we could get caught in inconsistent statements, and the groups to whom we wanted to communicate would find it hard to absorb our message. The best description of our message comes in what we call our Purpose and Aspirations. We have mentioned it already but it bears repeating:

We want to discover, develop and successfully market innovative products to cure diseases, to ease suffering and to enhance the quality of life. We also want to provide a shareholder return that reflects outstanding performance and to adequately reward those who invest ideas and work in our company.

It is no accident that the patients come first in this brief statement of Company goals. Yet, there is no attempt to conceal the fact that the company is set up to make money. Making money is not the company's only purpose, but it is a vital part of why the company exists.

As we began to formulate what we wanted to say about Gleevec—and to decide when to say it—the first question we asked was "What is the purpose of any pharmaceutical company in society?"

We knew that it was to bring better medicine to patients and to provide better tools for physicians—one and the same thing.

It makes no sense for a pharmaceutical company to boast that its drugs are profitable. Patients are not interested in that message, nor for that matter are many others.

The task of what to say about Gleevec became easier because everyone likes to read about major medical breakthroughs. We all know of someone who had cancer and know that one day we might become a victim of this disease.

The tendency with medical breakthroughs, of course, is to over-hype the message.

For so long, scientists have toiled in laboratories producing no news whatsoever. When they produce a successful drug, the pharmaceutical company will understandably feel a sigh of relief. And the company's marketing people will realize they have a very good thing going for them—an easy sell, as it were. They might beat the drums and find all the right superlatives to hawk the product.

That is when the trouble can begin.

A product like Gleevec looks so promising that it is hard to resist the temptation to describe it as miracle, or wonder, drug. And the media will often act as a willing partner. Such breakthroughs are rare and when they occur, they are dramatic so people want to hear about them.

Increasingly, based on better knowledge and understanding, the media and the public, brighter than ever, are curious about possible new breakthroughs, but they are at the same time skeptical or even cynical about any hype that might surround the publicizing of a drug.

Along comes Gleevec.

While Novartis exercised a good deal of caution in releasing information about Gleevec in its early stages, a growing fraternity of Gleevec users began to compare experiences—via the Internet—with the drug. They also lobbied Novartis to increase the drug's overall supply.

What did Novartis decide to say about Gleevec?

We pointed out that the drug appeared to be a breakthrough; that scientifically, it represented a new paradigm in cancer therapy.

We also made clear that, if the data were to be confirmed, the breakthrough would benefit only a small population of patients. The result for Novartis: it would not stand to make a large financial gain.

We were always careful to say that we did not know if Gleevec would be a cure for cancer patients suffering from CML. Nor did we suggest that the drug might work on other forms of cancer. We did not try to suggest that the drug, once it gained FDA approval, would be cheap and affordable for everyone. And we did not say the therapy would be short or that Gleevec produced no side effects.

But we did say that the drug's efficacy was better than any therapy we had seen for CML; and that the side effects were milder than those of other cancer therapies.

We emphasized that patients would have to take the pill for a long time, maybe for life. And, while Gleevec might end up curing CML, it was far too early to know or to say.

Other Cancers

We were careful not to suggest that Gleevec might work on other cancers. It was simply too early to know.

For the most part, the arrival of Gleevec on the American market received glowing media coverage, especially at the time of the FDA's approval in May 2001. The coverage was widespread as major television networks, radio, and newspapers covered the story. The media accepted Novartis's contention that Gleevec was a breakthrough drug in cancer therapy.

Looking back at the massive and positive media coverage the week after FDA approval, we were obviously delighted.

Part of the explanation for that kind of coverage had to do with the way many in the media covered the story. Rather than rely solely on Novartis news releases, quite a number of journalists sought out CML patients taking Gleevec and included their stories in their coverage. Nothing was more credible than these patients telling their own stories.

What was so remarkable and pleasing was to find a near absence of cynicism toward Gleevec among the media. It was not that the media believed, or tried to suggest, that the drug was indeed a cure for all forms of cancer. But no one did stories that suggested it was too early to tell whether Gleevec actually worked on patients. Interviews with CML patients whose lives had been changed radically after taking the pills put to rest any cynicism that might have existed among the media.

It was inevitable perhaps that all that positive coverage would put Gleevec on a pedestal. However cautious and careful we at Novartis tried to be, and however accurate the media was in describing the limitations of the drug, all that glowing media coverage left the impression that Gleevec had no defects whatsoever. But such coverage proved fleeting.

It did not take the media long to change its collective views of what constituted the news about Gleevec: at the time of the FDA approval, the news was of medical breakthroughs and startling recoveries. Just six weeks later the media defined the news surrounding Gleevec as anything that cast doubts on its efficacy.

Stories began to appear that were really not new, but because "old news" suddenly showed up in a scientific journal, a news story became justified. That was the case when *Science Magazine* wrote about two mechanisms of resistance to Gleevec in blast crisis; the *Wall Street Journal,* using the piece as a peg, wrote a story on June 22, 2001 headlined "Gleevec Shows a Weakness in Fighting Advanced Cancer: New Study Details How Disease Resists Drug" about those findings, giving the impression that Gleevec had defects, that it was not the breakthrough drug that we had promoted it to be.

While we firmly believed that the *Science Magazine* piece was just the publication of news already known in the scientific community, the *Wall Street Journal* article had the effect of deflating some of the optimism and excitement surrounding Gleevec. What could we have expected from a story that led off: "Gleevec, the cancer therapy hailed as a wonder drug against certain types of tumors, turns out to have an Achilles heel after all . . ."? It went on to report that 80 percent of late-stage patient, with CML, who initially responded to the drug, see their cancer return within six months, an often-fatal relapse.

The problem with the article was that it appeared to suggest that Gleevec offered patients a great initial response, but then in time the effect of the drug on the patient wears off. We have tried to make clear that as of now, the effect of Gleevec on patients in the advanced stages of CML, who are after all very ill, is limited. But it is worth recalling that these patients did not respond to other treatments either. Nor is it a secret that in the most advanced stages of the disease patients do relapse after Gleevec therapy; some 20 percent however continue to respond.

In the general excitement surrounding Gleevec's efficacy, the fact that it did not work for everyone or that it did not work forever in advanced cases was lost. The *Wall Street Journal* article helped remind us that we needed to do a better job.

The point we would have wanted to make much earlier to journalists who began looking for "warts" in Gleevec was that nothing that the *Science Magazine* article mentioned lowered the value of the drug.

Managing Expectations

Certainly the media will never treat Gleevec as glowingly as it did at the time of the FDA approval but nevertheless its therapeutic value has been confirmed, as well as its superiority to existing treatments, especially when therapy is indicated early during the disease. One critical task we at Novartis will have will be to manage expectations about the drug as information comes out about its effect on other forms of cancer.

We are quite conscious that Gleevec will perhaps not have the desired effect on cancer other than CML and GIST. We will have to make sure that the public understands that the value of Gleevec is not diminished by these ups and downs. The true value of Gleevec is this: not only can the lives of those with CML or GIST be prolonged but also the drug may well become an important new paradigm for the treatment of cancer therapy. Accordingly, we will hopefully find that many new cancer-fighting drugs hit the market based on the approach used in the creation of Gleevec.

≋

The Gleevec story has given an undeniable boost to Novartis. Here was a concrete example of this company's commitment to innovation and a tangible example of how its cancer research had become among the best in the world.

The development of Gleevec also demonstrated that the company could rally around a product in the interests of the patient (let's never forget how few patients Gleevec targeted at first) and that the management was willing to take some risks and place some bets in order to move forward quickly even if Gleevec was not very commercially attractive. The patient assistance program that Novartis organized for Gleevec patients also showed that the company was not insensitive to the difficulties of some less fortunate patients who did not have the means to acquire the drug.

We have been so pleased to recount the story of Gleevec because the drug is a perfect example of what Novartis stands for: alignment within the company, caring for the patient, an emphasis on innovation, our ability to improve the quality of life, to prolong life, perhaps even to cure a deadly disease. Gleevec has given us the chance to communicate these attributes of the company internally and externally.

≋

Success, as the expression goes, has many fathers.

And when Gleevec's success became apparent, many wanted to share in the glory.

Conscious even at an early stage that it would be important to clarify who contributed to the discovery and development of the drug, we wanted to avoid confusion, quarrels, and injustice. At first, we planned to make these clarifications internally; we wanted those clarifications to be fair, clear, and transparent.

When people dispute who should get credit for the discovery of a drug, nothing good can emerge. I was witness to such a disagreement when at Sandoz and it was not pleasant. It was always my intention, should circumstances ever warrant, to avoid a repetition of that incident.

Now Gleevec had arrived and the "fathers" would soon be taking credit. To clearly state the people at Novartis who deserved credit for the discovery and development of Gleevec, we first created a list and then, along with some colleagues, evaluated where people fit among certain categories. We selected three categories—crucial, helpful, and adjacent.

Ultimately, after making it clear that the discovery was the result of many people's work, I was pleased that the media gave credit to Alex Matter and his Novartis team for their work in discovering this drug. Others outside of Novartis may feel that they contributed in one way or another to the drug, and some made enormous contributions. But as Alex and his colleagues discovered Gleevec and were involved in its development from the very start, it is fair that the focus remains on them.

≋

Even as we parceled out credit for the discovery of this breakthrough drug, we were all too aware how early it was in the whole process, early enough for skeptics—and there are enough of them around—to wonder what all the fuss was about. They had read the reports in the science journals, noted the immense media coverage— the *Time* magazine cover, the extensive television coverage, the reams of copy filed by science and medicine reporters—and yet they intimated that Gleevec was no more than a passing fad.

American culture, the cynics argue, likes a good life-and-death story, its media gets high on stories of wonder drugs and even higher when the drug starts to show defects and perhaps is even pulled from the market.

You watch, they continue to say, Gleevec will be just one more passing fad.

After all, how many wonder drugs come along in a generation? What nerve to say that Gleevec is a breakthrough drug. Why, patients have only been taking the capsules since June 1998.

And let's not forget, the cynics continue, the so-called spectacular results arising from those patient trials have not produced a cure for cancer. The results have not even demonstrated that patients will remain in remission for a very long time. About all those trials have shown is that Gleevec can give terminal patients a limited reprieve.

Is that enough to turn the world upside down?

And even if the capsule did work, the skeptics go on, only a very small number of people will benefit from its effects, so why all the shouting, why all the excitement?

≋

Of course the cynics may be partly right, as there are still several open questions about resistance and side effects. And our not yet having answers gives those cynics a field day, allowing them to drive a very large wedge between all of us who happen to be optimists—not wild-eyed optimists, just the plain old garden variety kind—about this drug.

Deep Optimism

I do have a deep optimism, but I like to call myself a realistic optimist as I always combine my optimism with realism.

I do not want to be disappointed nor do I want to disappoint others. But I must tell you that I disregard dreamers.

They dream. They do not *do*. They are as dangerous as the cynics. You have to be willing to try to achieve things.

So that is the dilemma.

At this writing—it is December 2002—it seems prudent to report on what we know about the drug, and what we do not.

What is it that we know for sure about Gleevec?

Certainly from the scientific perspective Gleevec is a breakthrough. To be very blunt about it: Some people who would be dead are alive. This is what I know. Also we have initiated one of the most comprehensive patient assistance programs, ensuring a broad access to Gleevec worldwide.

I also think Novartis will make some money from the drug.

I know that, with Gleevec now available, doctors are reevaluating how frequently to turn to the only documented cure for CML until now, a bone marrow transplant. These doctors are using the transplant procedure more selectively. In the past, if a CML patient had a suitable donor and was young enough to tolerate the procedure, the idea was to opt for the transplant immediately. Now the thinking is changing as doctors choose to rely more on Gleevec and less on the transplant.

What remains unknown?

I do not know if the effect of this drug will last because, as mentioned above, we have only seen the drug in action since June 1998. We have seen blast phase patients respond positively to Gleevec, but we have also witnessed their relapses, a sign that their cancers are less responsive to the drug.

There is always a risk that some cancer cells will mutate and develop an enzyme to destroy Gleevec. Cancer cells are very "intelligent"; they can develop resistance against a treatment and, again, we just do not know if the cancer cells might develop that resistance. So you have to kill them all and not give them a chance to develop any resistance to the drug; so early Gleevec treatment seems most appropriate.

But it is heartening that the scientists want to carry on the fight against CML and to try to understand the nature of this resistance. Hopefully, in the not too distant future, they will come up with Gleevec-like drugs, perhaps in combination with other drugs that will effectively combat that resistance.

It's been clear that CML and GIST, because they are caused by single mutations, are more likely to respond to medication than other diseases with many mutations. It's also been clear that Gleevec does not perform as well in circumstances where a disease has a multitude of mutations (as in the blast crisis of CML). Resistance to the drug develops quickly as the Bcr-Abl gene mutates in ways that allow it to get free of Gleevec's effects.

When might we know if one of our Gleevec patients is truly cured?

Normally we say that when a cancer patient is disease-free for five years and off the drug, we would call that patient cured.

That would be ideal.

What do I anticipate will happen?

In theory, one would say that these people are disease-free; they are cancer-cell free; but they are still under treatment and one can never control all bone marrow and blood cells. The question is: At what point would we say "Let's take them off the drug and observe them closely." That will be a very important step for us to take.

I believe there's a good chance that some of these patients are cured; but I do not know.

Alex Matter would be quite satisfied if cancer patients could reach their normal life expectancy with the cancer still in them, and be treated with well tolerated drugs: "That for me would be a perfectly worthwhile goal, that cancer becomes a type of manageable disease such as rheumatoid arthritis—but that people reach a normal life span. This would qualify as a kind of cure for me. But with Gleevec it's much too early to say."

Deep down, I have the soul of a skeptic. I tend to believe that things are never as bad as they look, or as good as they look. The best approach is to say: Let's see what time tells us. We have had over 8,000 patients in the clinical trials for CML. After 10,000 patients have taken the drug for a few years, we will be able to say a good deal about it.

What we can say for sure at this stage is that Gleevec is absolutely revolutionary. It does more than we thought it would do. It's changing people's lives. It will change the practice of medicine in treating two diseases of today.

We are pleased to talk about the special qualities of this drug, not just to promote our company—although we *are* delighted that it is Novartis that has researched and produced this drug—but because, as people get to realize how much we believe in this drug, they will begin to use it.

Why the Fuss?

Why was there such a fuss about Gleevec?

After all, other breakthrough drugs have surfaced over the years, but none received the massive media attention this drug did; an American cabinet member had never called a news conference to announce FDA approval for any other drug. Research scientists and physicians alike have never rallied behind a drug with such public enthusiasm as did those who were Gleevec's official cheerleaders from the day of its FDA approval. Company employees rarely give up their weekends and vacations to help expedite a product's journey to market, as was the case with Novartis personnel.

Why did the media demonstrate almost universal praise and awe for this drug at the time of its "victory in May"? Usually quick to join critics, normally proud of its fiery independence, the media this time joined Gleevec's other cheerleaders, and happily proclaimed this tiny orange capsule a "magic bullet" of sorts.

No editor of a major news organization thought it too early to do in-depth stories on the drug. In those first days after FDA approval, no editor thought it worthwhile to criticize Department of Health and Human Services Secretary Tommy Thompson or Acting FDA Commissioner Bernard Schwetz for parading Gleevec in public as a drug that could well be the breakthrough drug we have all been waiting for in cancer therapy.

Why all the fuss?

Plain and simple: the early results were nothing short of spectacular and the mechanism of action opens new horizons of cancer-targeted treatment.

Still, none of us at Novartis could have predicted such an outpouring of goodwill and enthusiasm for Gleevec as occurred in those early days after FDA approval in May 2001. Of course, from the very first hours that we saw the stunning results of the Phase I patient trials in April 1999, we began to believe that STI571 might have some very interesting positive effects on cancer patients.

And now that we have all witnessed the outpouring of excitement for the drug, expectations for its future have grown dramatically. And that is not surprising either. Gleevec did not come to life in the dark shadows; it was swept to prominence by a giant wave of media exposure that increasingly exposed all of us to questions about its possibilities.

We were asked to read the future. We were asked to say precisely how this drug would work on other cancers.

And of course we could not. We could only talk about the first four years of Gleevec and even then we felt an obligation to be exceedingly cautious. We were barraged with the same question: Which of the other cancers will Gleevec help? We had to say that we did not know, that we were intensely studying the question.

≡

We were aware that Gleevec had an effect on a very limited number of cancer patients. We would have liked nothing better than to find that the pill worked on other cancers that affected far larger numbers than CML or GIST.

And so we have asked our laboratory scientists to devote their time to the question of Gleevec's future.

They are looking into Gleevec's effects on cancers that affect the breast, the prostate, the brain, the colon, and the lungs.

Some scientists have speculated already that the very nature of Gleevec—that it targets a specific molecule—means it will not work on other forms of cancer; that it will require other drugs specific to the molecule causing those other cancers.

Such thinking makes some sense. But again, we do not know.

Noting the kind of media attention Gleevec has already received, we have been very careful, needless to say, in divulging any information about the studies we are sponsoring into Gleevec's effects on other cancers.

We have learned that we must be careful in how we divulge that data. We always want to be truthful. It is very important not to be disingenuous.

The truth catches up with you anyway.

In the short term you lose a great deal of credibility. As long as you tell the truth, at least you have been honest, even if what you believed about the efficacy of a drug turns out to be wrong.

We want to wait until we have definitive results because we know that such data will receive widespread attention. If Gleevec works on even one of these forms of cancer, or even if it works in conjunction with some other drug, the efficacy of Gleevec will be magnified many times over.

Already physicians and investigators are speculating about our studies on other forms of cancer. One investigator assumed that because we had not published any findings on these studies by the summer of 2001, it was safe to say that we had nothing positive to announce. He based his hypothesis on the fact that we had been quick to publicize the positive findings involving Gleevec and GIST.

The truth is that we are trying to manage the news about Gleevec's future carefully and there is no reason to read into our silence on this subject too much.

A Gleevec Cocktail?

Some scientists are looking into the question of whether Gleevec might work in conjunction with other drugs. The argument is that Gleevec alone might not be effective on certain kinds of cancers; but taken as part of a "cocktail" of drugs, it might help patients. We can take heart from past cancer therapies that almost always include some combination of drugs. There is almost no cancer treatment based on a single drug. We certainly need to explore the "cocktail" approach.

Professor John M. Goldman, for instance, began a Phase I trial in

July 2001 on Gleevec in conjunction with interferon-alpha. Dr. Brian Druker wants to undertake a three-pronged study looking at Gleevec with three different drug combinations.

Our own Alex Matter is interested in exploring how natural compounds, such as fungi and plants, can be used to develop new cancer drugs. One promising compound in early stage testing is called Epithilone B, a fungus that is proving effective in interrupting cellular growth—a process that can ultimately kill tumor cells.

Under Alex Matter's and Greg Burke's supervision, Novartis has ten cancer compounds under development. Over the past five years, we have gone from having almost no influence on oncology treatment to being one of the leading pharmaceutical companies in this field.

For Alex Matter, the true significance of Gleevec is that it changed the mood of the research community: "They see now after all that we may not do these things just for fun. It has changed the mood of the pharmaceutical industry: People in the industry have begun to think, 'Yes, we can make innovative breakthrough therapies and be successful.'" And the mood of the regulatory people has changed as has the mood of the patient communities: After Gleevec, Matter feels, the general public has a different perception of what we do—a slightly more positive one: "At least we don't hear every day that animal experiments are horrible; that gene technology is for the birds; and that all of these drugs are doing nothing, and whenever drug companies say something, it's a lie."

Matter credits the change in the mood to the "missionary zeal of Brian Druker and his communication skills: He really inspired the clinical community, the regulatory industry; he was inspiring; he was a fantastic messenger and ambassador for this compound."

Could Gleevec become a paradigm drug, one that would serve as a model for other drugs using the same molecularly targeted approach?

Certainly, what Gleevec has taught us is that if we understand the specific abnormality that drives the growth of a cancer, we can then target the abnormality and develop an effective therapy with minimal toxicity. But it is essential that we understand the abnormality that drives the growth of a cancer. In effect, we are trying to figure out what part is broken, and then replace the part. And that will require a huge amount of work.

The problem is that so far in most cancers we simply do not understand the critical abnormalities. That is why the project to map genes is so important for it will help us identify the abnormalities. In effect, if we have the entire gene sequence, it is the same as having a complete list of our body's parts. We still need to know what is broken and the connection of that abnormality to a certain disease, but having the genome in hand will accelerate our work. As for any quick breakthroughs, we simply can't say. It will take years to uncover the function of the various genes. We may find the answer to some cancers in a year or two; others may take 20 or 30 years. At least we know, thanks to Gleevec, that we are on the right track. We know that we will reach success in other forms of cancer; it' s no longer a matter of "if," but "when."

≋

What of the future for the scientists who took part in the development of Gleevec?

All of our scientists take great satisfaction at being in on the creation of Gleevec. One, however, has become frustrated.

In November 2001 we came across Jürg Zimmermann sitting in his office pondering the future. He acknowledged that Gleevec to him was interesting but it had its negative aspects. "When you have a goal in your life, and suddenly you reach the goal, it can be very positive but it can also be frustrating. It creates a kind of emptiness because the likelihood that I will find a second "Gleevec" is zero, or almost zero."

Why such gloom about finding more "Gleevecs" in the future?

Because, he insisted, luck was involved in developing the drug at almost every turn.

"In so many times in the Gleevec story, I have seen nothing but luck. There was no logic. It was not that from A followed B and C. It was good luck that we had a merger. It was good luck that Alex Matter was a very good manager who pushed it; that Nick Lydon was working in biology. It was critical that we found Brian Druker. And so much of my own work was luck. We had to try many, many times when we thought we had it, and it was wrong. It's not like you can say, 'If I'm very clever, I will have it one day.' You need more than that: luck, and persistence. You continue to work and you struggle so many times. You have to stand up again and make the next molecule. And then it's toxic; oh no. Okay, try another one. Then it's metabolically not stable. It's so difficult really to make a molecule that eventually is a drug. It's really difficult."

But of course it was not just luck. Jürg Zimmermann suggests, in his case at least, there was a good deal of naiveté behind things. When he began the project, he thought it would not be that difficult: "I came from the university so I thought: give me a problem and I'm going to solve it. In academic life, you get a problem and you try to solve it. You write a paper. I thought it's the same here in the company. You are assigned a project; you do your job, and then go to the next one. Now when you look backwards you think: it's not so easy to do your job. A lot of lucky circumstances have to be in place."

Zimmermann feels badly for scientists who are forced to retire without being part of a success story: "That is why I said: you need naiveté to move things forward usually. As soon as you look around and see others who you thought were brilliant chemists, but they failed to do anything, if you start to think about that you might miss the drive you need to push your own work. As soon as you reach the status where you don't push things, then you should immediately leave the job. We have too many people in research who are like government employees."

One thing we acknowledge openly: We do not know what the long-term therapy prospects of Gleevec are, long-term in this case meaning the next four or five years. We do not know for how much longer patients will respond to the drug. Nor do we know whether we might be able to take patients off of Gleevec at some point.

So, what of the future?

We know already that Gleevec will not work in all cancers. But we hope that there might also be interactions between Gleevec and other yet-unknown drugs.

We also are quite aware that some of these developments will be treated as negative in the eyes of the media. We at Novartis will do our best to explain why we think they are not negative, or at least less negative than what the media is saying.

In our hearts, we know that Gleevec remains one of the best cancer therapies.

≋

For all our delight in Gleevec, we at Novartis want to make sure not to become known as a one-product company. A very important question for us is: How will we transition from one product to another?

The real question to ask is not: What is your most popular drug? Rather, it is: What does the company stand for? A few years ago lifestyle drugs like Viagra were all the rage. The conventional wisdom was that a pharmaceutical company had to have lifestyle drugs in its portfolio. Novartis was actually criticized for not having any of these drugs; but we were never stung by the criticism. It seems so much more important to us to be involved in researching and developing life-saving drugs. That is why Gleevec is close to my heart.

≋

The treatment of cancer is shifting ground to a whole set of signal-disrupting compounds that might provide improvements for patients. Gleevec is not the only one. The great advantage of these compounds is that they appear to have a positive effect on cancers without patients having to go through the terrible effects of chemotherapeutic drugs.

A second drug called Tarceva, being developed by OSI Pharmaceuticals of Uniondale, New York, and Genentech Inc. of San Francisco, and Roche Holdings of Basel, shows activity against tumors in patients with cancers of the head and neck, and two other forms of the disease.

AstraZeneca PLC is conducting a large patient trial on the drug Iressa, which has shown a good response in a small study of lung cancer patients.

Genentech's new breast cancer drug Herceptin was among the first such signal blockers to arrive on the market.

If these new drugs prove effective, they might turn cancer into a chronic disease that patients can keep under control for years. As in the case of Gleevec, cancer patients would take a few pills each morning that would manage their disease; heart patients and prospective heart patients do this now to control blood pressure and cholesterol levels.

There is no one cancer and no one treatment for cancer. There are hundreds of cancers and every battle which we win, is a battle, and not the war, and we have to go step by step, gaining ground, one after the other. There are major steps, and there are minor steps, and after a major one, there will be many minor ones.

What Gleevec shows is that one kind of a new approach based on microbiology and the newest discoveries of biology does work and opens new treatments of cancer.

And so Alex Matter and his team are still in the laboratories, waging other battles, trying to learn more about Gleevec, trying to come up with new drugs that will combat other forms of cancer.

Patients, physicians, clinical investigators—all of them—await eagerly their work.

The men and women in the laboratories try not to feel the pressure. But they know it is there—all the time.

The pressure does not change the reality. And the reality is that the race continues.

Afterword

It is November 2002, and I am leaping forward from April 1999 to the present. Though it is still very early in the story of Gleevec, we certainly can assess its contribution thus far.

For one thing, the drug has been prolonging life for many of the most severely ill CML patients. In some cases, it has rescued CML patients who were near death, giving them back a reasonably normal life. But what may be the most important finding is the fact that starting with the Gleevec treatment early in the course of the disease results in the best therapeutic outcome, as the drug has a greater and probably longer lasting efficacy. It certainly acts faster and better than any other available drug.

On February 28, 2002, *The New England Journal of Medicine* published the results of a study involving 454 patients with late-chronic phase CML; previous interferon therapy had failed with these patients. It showed that after 18 months of Gleevec, CML had not progressed to the accelerated or blast phases in 89 percent of the patients, and 95 percent of the patients were still alive. Patients now take 400 to 600 milligrams of Gleevec. The first-line CML data demonstrated that earlier use of Gleevec equated to a higher percentage of patients receiving a major cytogenetic response. In fact, in 68 percent of patients there was a complete cytogenetic response—that is there were no traces of the Philadelphia chromosome (the hallmark of the disease). This compares to the combination arm of interferon-alpha and Ara-C, in which only 7 percent of patients experienced a complete cytogenetic response.

Well, then is it a cure for patients with CML?

It is still too early to say. We cannot yet say for sure whether Gleevec is a cure for some patients with CML or any other cancer. While the results are startling in many patients, the maximum duration of treatment up to now is only three and a half years. We do not know if patients will still be free of cancer cells after five years and will be able to stop taking Gleevec. This would be a condition for really speaking about a cure.

Since 1999, physicians have tried Gleevec in a number of cancers other than CML. The most striking success was achieved with patients suffering from gastrointestinal stromal tumors (GIST). In May 2002, I shared the very moving experience of attending the first meeting of the Life Raft Group, a GIST patient organization. The meeting, which took place in a hotel in Cambridge, Massachusetts, included about 100 GIST patients, who were meeting for the first time. It was a deeply moving experience for all patients and their families, not just for me and Dr. George Demetri, who had treated most of them. Many of the attending patients would not have been alive without Gleevec.

On August 15, 2002, *The New England Journal of Medicine* published key data on Gleevec for the third time. In this issue, the data published was from 147 patients with advanced cases of GIST. The results showed that Gleevec induced a sustained objective response in more than half of these patients.

By September 2002, Gleevec had been launched in over 80 countries for CML and in all major markets for GIST. At the same time, we received a positive opinion from the Committee for Proprietary Medicinal Products (CPMP) in the European Union for Gleevec. On December 23, 2002, Novartis announced that Gleevec had received approval from the FDA for first-line treatment of CML. This is all very rewarding, but there might even be the potential for Gleevec to act in additional cancers and cancer-like diseases. The latest news is that Gleevec might help patients with polycytemia vera, a disease characterized by an uncontrolled increase of the red blood cells.

We are also expanding our patient support program to additional countries. In this respect, we were able to depend on the Max foundation, which is evaluating each request for patient support and has already done a great job in Latin America.

Last but not least, Gleevec is on its way to becoming a commercial success, which nobody expected. So I am tempted to say that the rewards come to the people who do the right things, irrespective of any short-term issues. For my part I am deeply grateful for having had the chance to be associated with this drug and a great team of people who discovered, developed, produced and commercialized it.

Let me close by saying that this story is not yet at its end.

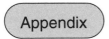

Patient Case Studies

Now that we have told the story of the development and production of Gleevec, we want to conclude with the human side of our tale. In this Appendix, we focus on a number of patients whose case studies illuminate, in one way or another, the excitement and drama of these tiny orange pills.

In acting so aggressively to develop and market Gleevec, we believed in our guts and our hearts that this was a scientific breakthrough medicine for a number of important reasons. In and of itself, Gleevec could help thousands of patients who had CML. But beyond that, we hoped that it would ignite much more research and lead to more drugs using the unique approach targeting molecular pathways.

We have had many wonderful experiences with CML patients. It has been icing on the cake to find that Gleevec helped patients with other forms of cancer as well.

Here is the story of one such patient.

Anita Scherzer

The GIST Patient

The doctors who were taking care of Anita Scherzer had a certain idea of what was wrong with her. They did everything under the sun to help her. But for a good long while after she was diagnosed with cancer, the doctors were battling the wrong form of the disease!

Anita turned 61 years old on October 11, 2002. For much of her life she was a pretty healthy person. Born in the Bronx, she graduated from high school and spent many years as a bookkeeper. In her same tone, Anita recounted the long and harrowing tale of her illness. "I've had cancer recur six times. I've had five different operations." As we'll see, she actually miscounted. She's had nine operations.

Renewing Her Vows

When married couples decide to renew their vows, they normally pick a round-numbered year. Anita and her husband Norman chose their 39th wedding anniversary on June 24, 2000, rather than the 40th to go through that pleasant task. "The thinking was that I wouldn't be around for the 40th," Anita observed stoically.

Afterwards, when she took a turn for the worse, she promised her grandchildren that for her 60th birthday she planned to take them all to Disney World in Florida. All 11 members of the family made that trip in November 2001, Anita included.

The renewal of the wedding vows, the trip to Disney World—these were Anita Scherzer's way of living every day to the hilt. She felt she had no choice. It had been seven years since she was diagnosed with cancer.

Only an Ulcer

In April 1994 she was having trouble breathing and believed that she had asthma. She went to a doctor for blood tests. Anita's hemoglobin was down to 4 (the normal low end of the range is 12). The doctor put Anita into the hospital.

Though the doctors originally suspected she was suffering from an ulcer, tests indicated that Anita had stomach cancer. The pathology report diagnosed the cancer as leiomyosarcoma. Surgery was performed: she had half of her stomach and her spleen removed. She then began six months of chemotherapy, once every three weeks. She held her own until November 1996 when her cancer metastasized to the liver. She had a second operation to remove her gall bladder and part of her liver.

Anita then had a spell of good health until early 1998.

She was getting pains in her right side near the hip. A CT scan that March, however, showed that she had a third recurrence of cancer, and surgeons at Memorial Sloan-Kettering Hospital in New York removed part of her ileac bone and right gluteus muscle, leaving her with a disabling limp.

The very next month—April 1998—she was again hospitalized, this time for a staph infection, which turned into an abscess and in June 1998, she went through four emergency operations in a four-week period.

The following November her cancer recurred in the same pelvic area and again she had surgery. In August 1999, again her cancer recurred, this time on her right adrenal gland; surgeons removed that—she had now gone through nine operations (five for cancer and four to close an abscess which developed from one of the cancer surgeries) since her cancer was diagnosed over five years earlier.

In April 2000, Anita and Norman were sitting in the oncologist's office at Mount Sinai Medical Center in New York City. Anita had just finished her CT scan and it showed that her cancer had returned for the sixth time. With a new tumor in her stomach, she was having trouble keeping food down.

Doctors asserted that the disease was inoperable because it had spread to too many places. A senior oncologist at Mount Sinai Medical Center proposed treating her with ifosfamide, a kind of chemotherapy; he based this recommendation on what he knew about soft-tissue sarcomas, the broader category of cancer to which leiomyosarcoma was a subdiagnosis.

Anita's husband Norman was a Disease Management Specialist for the Centers for Disease Control from 1961 to 1986 in Atlanta,

Georgia and from 1985 to 1990 he was Assistant Commissioner of Health for the City of New York.

Norman mentioned to the doctor that the literature he had reviewed showed that the chemotherapy proposed was partially effective against some soft tissue sarcomas, but not against the type his wife was diagnosed with, leiomyosarcoma. At that point Norman began to research the Internet for treatment options. At an on-line discussion group for leiomyosarcoma patients, run by the Association of On-Line Cancer Resources, he heard the term GIST used by a leiomyosarcoma patient. The patient also mentioned that a sarcoma specialist at Columbia Presbyterian had just determined that he (the patient) actually had a form of soft tissue sarcoma called GIST (Gastrointestinal Stromal Tumor) and that this kind of cancer was even more rare than leiomyosarcoma. The bad news was that GIST was much more resistant to all known forms of chemotherapy than leiomyosarcoma.

Fast-forward a few months. Anita and Norman were visiting the Columbia Presbyterian Hospital where a young sarcoma specialist, Dr. Mary Louise Keohan, was tuning in to the issue of GIST. She ordered a new pathology test for an enzyme called c-Kit. A positive report for c-Kit confirmed that Anita actually had GIST.

Dr. Keohan mentioned a new drug called STI571 that had shown some promise in the laboratory against GIST. The drug had been in trials, with significant success against CML, but the drug was not yet available for GIST.

Ready for Ireland

Norman began searching for ways to obtain the drug, discovering that Novartis was producing it in Ireland. Norman began thinking about traveling to Ireland to try to get the drug. How he would do so, once he got there, was not at all clear. Fortunately he never had to make the trip.

A late evening telephone call from Dr. Keohan alerted Norman that in a few months there would be a new clinical trial for GIST using this new drug, with 30 slots in the U.S. (and six more in Finland); Dr. Keohan suggested that there might still be an opening at Fox Chase Cancer Center in Philadelphia. Norman reached Dr. Margaret von Merehn, the clinical trial physician there.

Anita began taking six Gleevec capsules a day on August 22, 2000 in Philadelphia. She was Number 22 of the 30 patients in the first clinical trials for GIST; this was a Phase II trial because investigators had skipped Phase I, having established safety levels with the CML trials.

On September 5, Anita had a follow-up biopsy of her liver. She recalled that "they wanted to stick a needle into my liver to draw out whatever they have to draw out but they couldn't find the part of the tumor that had been there on a pretest. It had begun to shrink. That's how we knew the capsule had started working after only ten days."

Thirteen days later, Anita went for a CT scan back home in New Jersey and she recalled, "The radiologist was ecstatic." The results were faxed the next day to Fox Chase: "The radiologist came out to my husband and asked him if I had had an operation for the cancer recently." Norman said she had not. What had happened was that her tumors had shrunk over 50 percent in less than thirty days and the remainder was beginning to dissolve into liquid.

A month later—October 18, 2000—Anita had another CT scan and the tumors had shrunk even more; everything that remained had liquefied. This CT scan was not required as part of the trial protocol, but the physicians were so astounded by the one-month CT scan results that they added this one to reconfirm the original CT scan.

Anita began to feel a bit more energetic. The Gleevec capsules carried side effects but she found them tolerable. The worst: her skin became thin and if she scratched herself, her skin broke; for a while her indigestion, which had bothered her prior to starting the trial, became much worse. She had some diarrhea some mornings but Imodium eased that. She found the period before 10:00 a.m. the most difficult.

Other than taking those six capsules each day, Anita's life regained a certain normalcy. She got out of the house more. She no longer felt sick.

Norman Scherzer found it fascinating ". . . that only a year and a half ago I was working under the proposition that Anita had leiomyosarcoma. To then discover that she actually had an even rarer cancer called GIST, which is diagnosed by a test to determine the presence of an enzyme called c-Kit, and then to discover that she had a mutation in exon 11 of a particular gene of a particular chromosome which was causing her cancer cells to grow uncontrollably and whose growth signal could literally be turned off by a new targeted drug, now that is pretty mind-boggling."

He was also impressed with how much progress Gleevec represented for cancer therapy: "Because this is a molecularly targeted drug, this is the difference between standing across the street and throwing a rock against the building hoping to hit a guy inside versus having a futuristic laser-guided raygun and shooting him in his earlobe. It is an advancement of science that is so incredible that it is the future of how every disease is going to be treated in the 21st century. That's the breakthrough with Gleevec. My own breakthrough is that my wife is alive as part of a scientific process—and once people understand this, you'll see lots of other drugs like this in the molecularly targeted category.

" I firmly believe that the discovery of Gleevec and its application as a cancer treatment, in the 21st century will become comparable in importance to the discovery of antibiotics, and its application to treating most infectious diseases, in the 20th century. Molecularly targeted drugs like Gleevec will become the model for treating cancers and other life threatening and crippling diseases."

Richard Rockefeller

The Doctor Patient

Practicing family medicine for 20 years, physician Richard Rockefeller never once had a patient with leukemia. His knowledge of the disease was mostly from medical school.

That is why when the symptoms began—while he was traveling for Doctors Without Borders in Uganda, he dismissed them as indicating something he had picked up there—and which he hoped would go away on its own!

Born in New York City on January 20, 1949, he practiced and taught family medicine in the Portland, Maine area for many years. He developed a strong interest in helping the world's disadvantaged acquire quality health care and served as the chairman of the advisory board of the American branch of Doctors Without Borders (in French, *Médecins Sans Frontières* or MSF). Richard Rockefeller and I have a personal connection. We met as members of the board of Rockefeller University in New York City.

He has spent much of the past decade working on various health and philanthropic interests, including a non-profit foundation he created—Health Commons Institute—which promotes information technologies that bring complete biomedical information to patients and doctors.

A Visit to Africa

In May 2000, Richard was traveling in Uganda on behalf of Doctors Without Borders, working on ways to improve treatment for sleeping sickness in central Africa. One day he found that his ankles had begun to swell; he was experiencing shortness of breath; he could hear his heart pounding in his ears; something seemed wrong. Perhaps, he thought, it was the heat, or something he ate. Maybe he had picked up a parasite.

A week later he returned to Maine. He usually jogged 15 miles a week. When he tried jogging he found that his energy had been sapped; and the pounding in his ears persisted. Still, since the swelling in his ankles had disappeared, he busied himself with his work, keeping his concerns to himself. He did stop jogging.

That September he and his wife, Nancy, were staying at their cottage on an island near Acadia National Park in Maine but the usual activities hiking, windsurfing, working in the woods brought on heavy fatigue. Returning home in October, after running only a quarter mile he felt like he was dying. Over the next two weeks he grew substantially worse. He felt overwhelming fatigue climbing a flight of stairs.

By now he was quite alarmed.

Finally, on October 19, he ordered his own lab tests.

That evening a lab technician phoned him at home. Technicians were under instructions to phone the patient's physician at once whenever the test results indicated something serious. In this case, Richard Rockefeller was doctor and patient—one and the same!

Richard picked up the phone.

"Dr. Rockefeller. We thought you'd want to know that your white count is 194,000."

He knew enough to understand that a white cell count of that magnitude could mean only one thing—leukemia! Unable to process the technician's words, he asked him to repeat what he had said twice more. Perhaps the technician had added one too many zeroes—how bad off would he be if his white cell count had been only 19,400? (As we've noted earlier, the normal range is between 4,000 and 10,000.)

But not a chance, the technician replied; he had even asked the head of the lab to take a close look at the slide.

Richard translated the technician's words into a possible death sentence: "You may die soon or at least your life will change forever."

A Time to Meditate

What should he do now?

Should he call Nancy? His children? Friends? A specialist? Should he go on the Internet? It seemed altogether rational that he might just panic and scream. Then he hit upon something that might calm him.

For 18 months he had been practicing a form of sitting meditation, an activity that brought him great peace of mind in dealing with life's ordinary concerns. He decided to test its worth in the face of a true calamity.

For the first 15 minutes he was filled with terror. Then, to his amazement the terror passed, and meditating for another 45 minutes gave him an inner peace. He never forgot those first moments of terror or the calming moments that followed, and the meditation supplied him with a way to cope that he would take advantage of over and over in the next ten months.

But there was the cold, hard reality of telling Nancy and his two children. He has never forgotten how comforting their support was.

A few hours after the technician's call, Richard phoned Dr. Tom Ervin, a friend and oncologist.

"Should I be in the hospital getting my white cell count lowered?" He feared that the elevated white cell count might induce an immediate stroke.

The oncologist assured him that there was no urgency. But he advised Richard to come around in the morning.

Running his own tests, Dr. Ervin confirmed the high white cell count. He also found that Richard's spleen was enlarged. The disclosure made Richard wince: Here he was, a doctor, and he had not even identified his own enlarged spleen.

Dr. Ervin told Richard that he probably had CML. Blood tests the next day showed a 95 percent likelihood that he had the disease.

The oncologist put him on a dual regimen of hydroxyurea and allopurinal; the latter designed to prevent gout.

In one of their first conversations, the oncologist mentioned that new drugs were coming along in the fight against leukemia. He did not mention Gleevec. The oncologist proposed that Richard start on interferon and consider a bone marrow transplant within a year. With five siblings (four sisters and a brother) he stood a nearly 75 percent chance of finding a suitable donor (two sisters turned out to be perfect matches).

Richard raised the issue of life expectancy.

The median survival without a bone marrow transplant, the oncologist noted, was five years. If the disease had been diagnosed early enough, he might have as much as eight or nine years without a transplant.

Richard chose Memorial Sloan-Kettering Hospital in New York for his treatment because it was relatively close and he had contacts there. Two days after being diagnosed, he had called his father, David Rockefeller, who was lunching with his brother Laurence, who knew a Dr. Paul Marks; Dr. Marks referred Richard to Dr. Stephen Nimer, whom Richard called the following Monday.

The CML diagnosis became definitive when the results of a bone marrow aspirate, ordered by Dr. Nimer, came back.

Coming in Piecemeal

As each day passed, Richard learned more and more about his disease and the range of treatments: "At first it was all coming in piecemeal. Now I can give lectures on the subject in my sleep." Dr. Nimer knew all about Gleevec and explained its basic mechanism to Richard. Soon thereafter, Richard was visiting the new president of Rockefeller University, Arnold Levine. On Levine's desk was a picture of the Gleevec molecule, a computer-generated rendition based on X-ray crystallography.

Eager to expand his knowledge of CML and Gleevec, Richard went on the Internet, joining the E-groups.com CML listserve. He was in close touch with many members of the site. He had breakfast in New York with Brian Druker. All the while Richard waited for his white cell count to normalize.

He looked into the possibility of joining a Gleevec patient trial for newly diagnosed CML patients but decided against it, in part because he first wanted to reduce the number of leukemia cells in his body, and to retain the option, once he did start Gleevec, of using whatever dose seemed most appropriate, rather than stick to the dose prescribed by the Gleevec trials.

A Drug from Paris

One treatment relied on a form of interferon called pegyllated interferon or PEG-interferon. That offered the prospect of reduced side effects (compared to the standard interferon-alpha that CML patients were routinely given) and once-a-week injections (rather than daily ones). It appeared to be, at the time, more effective than interferon-alpha in treating leukemia.

It has been estimated that, at diagnosis, the average patient has between 100 billion and one trillion leukemic cells in the body. Since resistance of cancer cells to a drug is based on probability—the more cells, the greater the chance of resistance—Richard hoped to reduce his chances of resistance to Gleevec by reducing the leukemia burden that he carried.

The "peg" of pegyllated stood for polyethylene glycol, a chemical that helped the interferon circulate in the body longer, providing more continuous exposure of the cells to be killed to the drug. The reason: the half-life of interferon-alpha in the body was a few hours; of PEG-interferon, one week.

Richard needed to take on the disease right away: He was 98 percent Philadelphia chromosome positive when he had a bone marrow test on November 9, 2000.

He began the pegyllated interferon treatment eleven days later, a month after he was diagnosed with leukemia. The FDA had not yet approved the drug, but it had been approved in several European countries. Richard arranged for friends visiting Paris to pick it up for him there.

The side effects of pegyllated interferon were no picnic: "It makes you initially feel like you have the flu, with fever and chills, and nasty muscle aches. That lasts only a few days. Those eventually go away. But fatigue sets in and gets worse; and I developed a ringing in my ears because it is toxic to the nervous system."

The drug made Richard ". . . very forgetful. People talk about having an 'interferon brain.' It makes you unclear, stupid, and eventually depressed. It's not a pleasant thing. I would do it in a heartbeat again to save my life, but still, it was very nice to stop it when Gleevec became available."

Despite the unpleasantness, Richard was able to tolerate high doses of PEG-interferon. As a result (and because he was lucky, he said) Richard achieved a 93 percent reduction in leukemic stem cells in his bone marrow by January 2001. He had to stop the drug then because the high dose had also caused his platelets to drop perilously low. Six weeks later he started again on a much lower dose. By the time of his next bone marrow test in April, he had actually lost a little ground: back to 20 percent leukemic cells in the marrow.

Had he continued to improve on the PEG-interferon, he might have remained on it, but he decided to switch to Gleevec when it became available. Interferon was a powerful but non-specific drug, effecting not only the leukemia but other body systems as well. Gleevec, by contrast, was designed to attack only one specific molecule and was unlikely to do much harm elsewhere in the body.

Feeling Fine All the Time

Quitting the PEG-interferon in early May when FDA approval of Gleevec appeared imminent, Richard traveled to the Utah wilderness for a solo hike. Upon his return, he heard some wonderful news. On May 10, the FDA approved Gleevec. Richard was thrilled: "I hit the timing just right." Friends and relatives e-mailed him, congratulating him and asking whether he had heard about the FDA action.

He began taking Gleevec on May 25, 15 days after the FDA approved the capsule. He simply took a trip to New York City and asked Dr. Steve Nimer to write him a prescription for the drug. Richard took four pills a day, 400 milligrams in total. He could have written his own prescription, but he preferred getting it from Dr. Nimer.

No great fan of interferon, Richard now had first-hand evidence why he could not wait to get off the drug. Once he began taking Gleevec he had no side effects whatsoever. He knew that others suffered weight gain, fatigue, bone pain and other side effects. But he had none. "I take my pills in the morning and I feel fine all the time. I'm back to normal. It might have been reasonable to stay with interferon but once you switch over to Gleevec you never want to go back."

In early June I saw Richard Rockefeller at a Rockefeller University Board meeting in New York and I asked him how he was doing. I knew that he had started on Gleevec in May and was curious about his progress.

"Not so good. I'm really sore."

This was indeed disappointing to hear.

I told him I was truly sorry to hear that. I assumed this was a Gleevec side effect.

Then he laughed and assured me it had nothing to do with Gleevec. "I've been running ten miles a week. That's what's making me stiff!"

In mid-July 2001, a FISH test showed that Richard was 99.2 percent Philadelphia chromosome negative. "I've gotten to where I want to be. I feel like a normal person." The following November a bone marrow quantitative PCR test showed that Richard's CML cells had decreased to a very low level.

He had begun taking Gleevec the previous May 25!

Suzan McNamara

Completing Her Story

Forever Suzan McNamara will be known in the Gleevec story as the woman who organized the petition drive. A few months after she sent us the petition, Suzan enrolled in a patient trial in Portland, Oregon on January 1, 2000. Three days later she took her first Gleevec capsule. She felt good, and she felt for the first time that she was going to live.

Monitored in the hospital constantly for the first day or so, Suzan had no reaction to the drug. She did not have to remain in bed and so she and some other CML patients talked among themselves and watched television. She was something of a celebrity to the others: "I had so many patients who came up to me in tears—and I still do—thanking me for what I had done. They were basically thanking me for saving their lives. At that point, I didn't even care if I lived or died because I felt as if I had done my good in the world."

Eventually, the pill began to work on her in a positive way. Her hair began to grow back. Her skin started to clear up. She was hungry again; indeed she became obsessed with eating. Her medical counts reflected her general feeling of wellness. When she had arrived in Portland, she had 90 percent cancerous cells in her bone marrow. Within three months, in early April the percentage had dropped to only 20 percent. In early July, she was down to two percent; after nine months, in October 2000, to zero.

Eleven months after starting Gleevec, in early December, she took the PCR test that checks 100,000 cells, far more comprehensive than the FISH test, which looks at only 200 cells. For the first time the PCR test was negative: All 100,000 cells tested were at zero. She had no cancer cells—neither in her bone marrow nor her blood: "That put me at a molecular remission; so basically I was in remission after 11 months on the drug. I went from being practically on my death bed in a year to living a normal, active life with so much hope for the future."

She took up full time studies in molecular biology in September 2000.

In late April 2001, she spoke to the Novartis sales force in Orlando, Florida, one of 15 patients on Gleevec who reported on what the drug had done for them. It was just ten days before the FDA gave its approval for the drug and Novartis was getting its sales team in gear. She found it remarkable that the company she had lobbied against now wanted her to speak on its behalf: "Most companies wouldn't want the public to hear about me."

She remained a minor celebrity. The English-language *Montreal Gazette* put her on the front page. She was the focus of a *People* magazine story in its August 6, 2001 edition.

Darlene Vaughan

The Miracle Patient

Darlene Vaughan had fallen ill in 1997. A year later, she was diagnosed with a rare cancer called a Stage 3 leiomyosarcoma. She went through extensive operations; went on experimental drugs; took chemotherapy; but her tumors kept growing.

During the fall of 2000, she wondered if she would survive much longer.

On October 20, Darlene's friend Anne Murphy showed her an e-mail from Norman Scherzer, then a member of an Internet listserve focusing on leiomyosarcoma. After Darlene's treatments for leiomyosarcoma had started at UCLA in January 1999, Anne had found this listserve and became a member, frequently updating Darlene. Norman had emerged as a key leader and supporter within this large group of leiomyosarcoma patients/caregivers.

He pleaded with everyone on the leiomyosarcoma listserve to redo their pathology tests to find out if, rather than leiomyosarcoma, they actually had GIST.

If they did, he said, they might be eligible to get into clinical trials for Gleevec.

Norman's e-mail stated that virtually everyone on the trial who had reported in had had good results to date (lots of early tumor shrinkage). He urged his e-mail friends not even to wait for retesting, but to try to get on the waiting lists for the trials.

That same day, Darlene called Laura Wilson, the senior administrator and nurse overseeing the OHSU STI571 clinical trial, and was put on the waiting list at the Oregon site pending Darlene's c-Kit test results. By early November Darlene got the results of her c-Kit testing: positive for GIST.

A Reason to Celebrate

However bizarre it might seem for a person to feel good about learning that she had a certain kind of cancer, Darlene wanted to celebrate. None of the traditional treatments to deal with leiomyosarcoma had any impact on her tumors, which were only getting worse and worse, larger and larger, and more numerous. She knew exactly why she was celebrating: a drug appeared to exist for GIST and it was being tested in patient trials.

Darlene hoped to enter one of the Gleevec trials within the coming weeks or early 2001, preferably at the Oregon Health and Science University because it was closest to L.A.

It was not to be.

November 2000 also marked Darlene's move into a small, furnished studio apartment in Westwood Village, near UCLA. She was growing weaker. She could do little more than get a few essentials for housekeeping, linens, kitchen items and a few groceries. She felt little joy and even the smallest task seemed difficult.

Late in December, feeling better, she drove to Mesa, Arizona on December 22 to visit her older brother Bob and his wife Sandie. Her younger sister Linda was to fly in from Florida for seven days to join in the family Christmas. Over Christmas Darlene began to feel very weak, napping most of the time, eating almost nothing.

Unbeknownst to Darlene, she was building up high levels of ammonia in her system. Getting on Gleevec seemed more important than ever.

Darlene spoke with Laura Wilson in Oregon several times late in December to check where she was on the waiting list. On January 5, 2001 Laura told her that she could begin the Phase II clinical trial at OHSU on January 18. Again, more rejoicing; now a date had been set to start the drug.

During this period Darlene called Dr. Rosen with the news of the trial in Oregon. He told her that he would have the drug at UCLA as well and planned to start a Phase III clinical trial on February 1. Dr. Rosen assured her that she could enter the UCLA trials. Even though this meant delaying her start on the capsules, Darlene felt it was preferable given the proximity to the UCLA clinic. She feared she was too weak to handle the travel back and forth to Portland.

The Illness Gets Worse

She returned to Los Angeles for her monthly clinic visit on January 23.

She became extremely ill in late January and during the first two weeks of February. Going for a CT scan on February 6 was an ordeal. Her mind grew more confused. She hallucinated that the wood grain on the door panel had turned into pictures; she thought she could detect Benjamin Franklin's face as well as some animals on the door.

She slept a great deal, hardly eating; she alarmed her friends who were unable to reach her.

Meanwhile, she was subsisting mostly on instant breakfasts that she mixed with water, and oranges. Her strength was sapped. She was not able to leave the apartment to shop or take the garbage out or do any mundane tasks. Both Anne and another friend, Lois Curry, were almost in panic as they realized Darlene could no longer be alone without regular assistance. Linda arrived from Florida and took over shopping, cooking and feeding Darlene. She helped her to the bathroom, and took her to the clinic.

Still hoping to start the Gleevec trial, Darlene had become ineligible—at least temporarily—because of her rising bilirubin levels. She knew that she was in a bind: she could not get access to the drug and her health was quickly fading. Going over her situation with Dr. Rosen, Darlene confessed to being tired of all the doctors and drugs and perhaps the time had come to just say that she was ready to come to terms with death.

When her bilirubin levels began climbing in early February, Dr. Rosen arranged for Darlene to have a stent inserted in a liver bile duct. His thinking was that the tumors were so enlarged they were squeezing off the ducts causing the bile to back up and impact liver function. Another doctor told Anne and Linda to plan for the worst. His assessment was that Darlene would not be alive for more than a few more weeks.

Weak and Spacey

Most of the last two weeks of February Darlene was doing very badly. She was semiconscious at times, suffering from a good deal of confusion and the effects of almost total liver function failure. She felt as if she were just slipping away. On February 27 one of Dr. Rosen's young medical fellows, Dr. Maria Delioukina, excitedly reported to Darlene that she had just gotten approval to get Gleevec through the National Cancer Institute on the basis of compassionate use. Dr. Delioukina had become involved with Darlene's case in recent months as she routinely assisted Dr. Rosen in clinic. She had been working to get this approval as it was apparent Darlene's failure made her ineligible for the Phase III trial now under way. Darlene wanted to be excited, but in truth, with her health so bad, she could not get up much enthusiasm, and probably did not comprehend that this meant she could now go on the drug. She was extremely "spacey" and weak at this point.

On March 1, Linda was panicking upon realizing how bad Darlene was; she called Anne who went to the clinic where she "borrowed" a wheelchair for Darlene to use; together Anne and Darlene's sister took Darlene to the hospital and got her admitted that afternoon.

A team of doctors began working on Darlene. Her blood counts had deteriorated so badly that she was given three blood transfusions over the next day or two. Darlene had no idea what the doctors were doing.

Darlene's brother Bob and wife Sandie arrived from Mesa, Arizona. When they first spoke with Dr. Rosen, he cautioned that Darlene had less than a 5 percent chance of surviving.

The stent was not helping, and liver function continued to deteriorate. Darlene remained in a semiconscious state, only vaguely aware as friends and relatives stopped by her hospital room—essentially to pay their last respects to her.

Darlene had pretty much reconciled herself to the notion that she would die. She was at peace with her fate. She had talked to God on a regular basis. She confided in those conversations that while she did not want to "leave," if that was what she was supposed to do, she was okay with that.

A Pink Glow

At some point over that weekend a chaplain came to visit Darlene. They talked briefly. He left a small Bible and a card.

On Sunday evening, Darlene experienced a strange sensation: She saw a pink glow in the room, blinking on and off intermittently. She had no explanation for the pink glow. But she did say to herself: You know what, I don't have to die tonight because for one thing, they will hook me up to life support and keep me going for another 48 hours; so at least I won't die tonight.

Finally, on March 5, Dr. Rosen prescribed Darlene's first dose of Gleevec. The moment passed quite uneventfully for her. She had been given so many different drugs in the past few days that she had no way of marking the moment.

But at some point she swallowed a 100-milligram capsule of Gleevec.

She was too out of it to dwell on the capsule. Nor did Dr. Rosen think much of the moment. He was busy telling Darlene's family that the end would come soon, probably within a week.

While Darlene did not remember taking Gleevec that afternoon, the next morning she woke up around 10:00 or 11:00 a.m. Her nephew's wife, Natalie, was sitting quietly at her bedside reading the newspaper. Darlene began talking with Natalie, speaking fairly rationally, and normally. They started to discuss Nat's pregnancy (four months along). She and Natalie visited for a couple of hours, and Anne stopped by as well. Natalie was amazed at Darlene's demeanor, and of course very happy. Anne thought that Darlene's condition was "surreal."

Eighteen Hours

An hour or two later Linda arrived with Bob and Sandie. By now Darlene sounded quite coherent as she chatted with them. She spoke normally, in complete control of her faculties.

It had been only 18 hours since she had taken Gleevec!

About mid-afternoon Dr. Rosen poked his head in the room. He took one look at Darlene and seemed shocked. He was blown away. He could not believe his eyes.

"You look fantastic." That was all he could say at first.

Then, collecting himself, he said, quite thrilled with what he was seeing: "This is miraculous."

Darlene stayed at the same 100-milligram dosage of Gleevec for several more weeks. It seemed remarkable that while others in the clinical trials were starting Gleevec at 400-milligram or 600-milligram doses, she had had this incredible response with only one-fourth the dosage.

Dr. Rosen explained to her that her body, in its weakened condition, would not have been able to tolerate a larger dose. Hence, she was given the smaller amount. It was unbelievable that it had worked that fast—small or large dose. In fact, there was concern that because Darlene's liver was in such bad shape, Gleevec, with effects still not clearly understood, might have hastened her death.

Darlene rebounded quickly after that. She left the hospital two days later on March 8. A week earlier she had been admitted to the hospital with the most miniscule chance in the world of leaving alive.

She visited the clinic frequently at first; her blood was taken twice a week for the first month. Her bilirubin and ammonia counts began coming down. At its highest, the bilirubin had reached 16.4 when the normal level is .02 to 1.5 mg percentage (milligrams per 100 ml of blood).

She began walking her dog every day. Her sister returned to Florida.

Since leaving the hospital in March, Darlene had her first CT scan on April 27, eight weeks after beginning Gleevec. The comparison was made with her last CT scan on February 6—a month earlier than when she began Gleevec: Of the three largest tumors in the liver and posterior to the spleen, there had been an aggregate decrease in size of 47 percent; some of the smaller tumors had disappeared.

Darlene saw a GI specialist on May 17 to have the liver stent removed. Though he believed that her recovery was a miracle, he remained cautious and wanted to check on her liver during the stent removal procedure. Upon checking, he told Anne: "I think Darlene has someone else's liver. I can't believe this is the same person as three months ago."

Sixteen weeks after starting the drug—on June 21, 2001—a CT scan showed another 12 percent decrease in the three large tumors; Darlene had experienced a total 54 percent reduction in the size of the tumors. Just as importantly, she was feeling terrific. Her appetite had returned; she gained weight.

On June 25 she wrote an e-mail to Norman Scherzer, thanking him:

> Your e-mail of 10-20-00 re: GIST prompted me to have the pathology done on my tumor tissue and get on Oregon's waiting list when I tested C-Kit positive. THANK YOU SO MUCH for getting that word out to folks! I wouldn't have known to pursue the alternate diagnosis at that point."
>
> Now I am living in L.A. temporarily and doing my uphill climb and assessing where my life is headed at this juncture. My two-month scans showed about a 50% aggregate reduction in the three largest liver and spleen tumors (in addition to a month of growth not on the scans from 2/6 to 3/5). My blood work is normal; I have had some side effects, but manageable. (I) am looking forward to remission at some point soon.

The thyroid imbalance was corrected and the effects of that imbalance resolved, including most of the severe fatigue.

By mid-July 2001, she pronounced herself recovered, although in subsequent months as Darlene went through ups and downs with the disease, she began to understand that there would be no full recovery; but hopefully there would be a return to somewhat of a normal life.

She had a sharper and sharper image of what she went through in the hospital in early March. She realized that she never did go into the Gleevec clinical trials, but got the drug on a compassionate use basis.

"I'm Dr. Rosen's private guinea pig," she jokes.

On December 17, 2001, Darlene provided an update: she spent four weeks in the fall driving "way too much" in her phrase and returned home with severe anemia; she remained in bed for the next four weeks and received blood transfusions in early December. She was feeling much better by the time of her update.

As of November 2002, Darlene had relocated to Arizona to live with her brother Bob and sister-in-law Sandie. She has been working out at a health club on a regular basis, vacationed in Hawaii the previous September, and was planning to move into a new home near her Arizona family in the near future. She expects to return to work within the next year.

Despite all her ups and downs, Darlene remains the miracle patient.

Marco Nese

Buying That Dog

Earlier in the book, we watched Marco Nese, the Swiss businessman from Basel, suffer through years of interferon for his CML. A relatively young man, he was just getting into his career in business and in academic life when the disease struck.

After a number of bouts with pneumonia in the first half of 2000 he was forced to quit working. So devastating were the side effects of the interferon that he truly felt his life was slipping away.

Then, in June of that year, Marco picked up a Swiss newspaper one day and read an article that would change his life.

The article had to do with a new drug known at the time as STI571. He learned from the article that the brand-new compound largely inhibited the enzyme that caused cancer cells to multiply too quickly, live too long, and invade other tissues. The article also hinted at startling improvements in the first patients who had been given the drug in clinical trials.

Marco could not tell from the article how quickly STI571 might become available for general use.

Marco put in a call to his doctor. He wanted to know what the doctor thought of the new drug. To his dismay, Marco received a cautious response. The doctor had seen too many new and promising drugs come and go. Marco would not be deterred. He went to the Internet and found other CML patients who had been using STI571. Remarkably, they were reporting few side effects from the new drug. He liked what he heard. If he had weathered the worst that interferon had to dish out, he felt confident that he could withstand whatever STI571 would throw his way.

He managed to take his MBA exams in July 2000, finding it extremely difficult to concentrate because of the pneumonia and high fever. He was highly motivated: he wanted to qualify for work as a consultant. He smiled at recalling the advice he got from others: His mother told him, given his deteriorating health, not to continue with the MBA program. But he was determined to pass the exams—and he did.

By now, Marco Nese believed that he was in a race for his life. His only hope seemed to be to enter one of the patient trials testing STI571. He told his doctor in great frustration that he could not go on anymore. He begged the doctor to arrange for him to go on STI571 at once, if possible. It would take a few more months. In April 2001 Marco Nese's physician instructed him to quit interferon as a prelude to entering patient trials.

For the next three weeks Marco took no drugs. Freed of the interferon, he grew stronger. He slept less each day and his demons had gone on holiday. Indeed, from the very first day that he stopped taking interferon, Marco senses a wonderful change for the better. He was not really getting better. He still had the CML and the "tiger" that was sleeping within him could arouse at any minute and consume him. He knew that. He was under no illusions.

Free of Pain

No matter how good he felt, he knew that his days remained numbered. As the time approached for him to begin taking STI571, he understood that there was no guarantee that the tiny orange pill would work. About all that Marco could hope for was that the pill would prolong his life, giving him the same ten years that interferon had promised—but with none of interferon's terrible side effects.

He knew he was taking a certain risk. Under interferon there seemed good reason to believe that he would last the ten years, not just five; going off of interferon meant that he might not live those ten extra years. STI571 might not work—and he had no desire to go back on interferon. He was playing for high stakes and he knew it. But he wanted to give the new drug a shot.

He looked upon April 24, 2001 as something of a red-letter day for him; that was the day that he began taking Gleevec.

Each morning he took four of the pills—each one, 100 milligrams. They came in a small plastic bottle. No matter what the pills did for him, they were certainly easier to take than the daily injections of interferon. He was grateful for that. He was to eat something before or after taking the pills, but he required no special diet otherwise. Just as when taking interferon, he was not permitted to smoke or have a drink. He followed those strictures—mostly; on occasion he smoked a cigar.

But the real question was whether Gleevec would truly help him get better. It was wonderful to be relieved of the side effects of interferon. He began taking the pills. He got through the first day, then the second, and the third. After a week he paid a visit to the hospital for his weekly check-up. The big concern of his doctor was that, in the weeks that he had gone off interferon, his white cell count might rise, a sign that the deadly CML was getting worse, not better.

Slowly, but steadily, his white cell count in his blood began to drop. Gleevec seemed to be working. That news alone should have made Marco Nese giddy with joy. But for him the most important thing was that Gleevec produced no side effects. For the first time in months, he found that he could concentrate; his memory was returning. Best of all, his fits of depression disappeared. He began to wonder what it might be like to live something of a normal life again.

Meanwhile, Marco Nese's hospital tests went very well. Gleevec was keeping his white cell count down. The cancer cells still remained in his bone marrow, but as long as the side effects remained tolerable, Marco was sure he could live a more or less normal life despite having CML.

A few days short of two months since Marco Nese began taking Gleevec, he traveled from his suburban Swiss home into Basel, to participate for a few hours on a panel about his experiences. I had tho privilege of looking into Marco Nese's eyes in June 2001 when he attended our Novartis "Family Day" celebration of Gleevec. He spoke movingly about what his life had been like before taking the drug— and he spoke as well about how his life had been since taking Gleevec. I felt so happy for him.

Doing Normal Things

The next morning Marco spent an hour and a half talking to my writer colleague Robert Slater about those experiences, sipping coffee in the restaurant of the Three Kings Hotel, overlooking the Rhine River. Pausing at times during the conversation, he simply gazed at the river and smiled, as if to say how truly wonderful it was to have the strength to hold such a talk and to be doing normal things. He pondered a return to work, and teaching seemed the most practical possibility because he would have more control over his time.

Meanwhile, he was thinking about getting that dog in the fall because he lived alone. But he was still cautious and unwilling to commit himself. His principle, he said proudly, was to do what he liked to do. With Gleevec, he had made a start on doing some of those things.

After taking Gleevec for two months, Marco felt hopeful. He dreamed that the drug would become a total cure; but he was entirely too aware that patient trials had been going on for just three years, far too short a time for doctors to pronounce the pill a true cure. He would just have to wait. The waiting brought tensions. He feared that the tiger might awaken while he took Gleevec. He kept telling himself that the doctors knew terribly little about the pills; they knew that its positive effects had lasted several years in patients; but there was no proof yet that Gleevec could keep someone alive four years, or five years, or six years, let alone the ten that seemed likely under interferon.

And yet Marco Nese felt excited.

Gleevec offered him new hope: the chance for a possible cure down the road; a recovery period that would be free of major side effects; and a therapy that required no more of him than to take a few pills each morning.

Early that summer of 2001 he was content to mark his progress day by day, planning some part-time work in the near future perhaps, hoping to buy a dog a few months down the line, thrilled to be motivated again to do things. He liked his new life, sleeping only ten hours a day, the disappearance of the depression. He liked taking a few pills each morning rather than getting interferon injections each day. Most of all, he could not quite believe what those pills had meant for him. At the very minimum, he had bought some precious time for himself and had enjoyed a vast improvement in the quality of his life. At a maximum, he might well be on the way to a complete recovery.

We caught up with Marco Nese in November 2001. Things were going pretty well, though he experienced periods when he felt stronger and when he felt weaker; periods when he felt more motivated and periods when he needed to rest. He was doing much better with Gleevec than he had with interferon. In September he went back to work part-time as a consultant at the Novartis Foundation for Sustainable Development that deals with health issues in developing countries. He was consulting on a malaria project. He was working from home, which he found the best way to manage his disease each day. Seeking to avoid stress, he did not want to rush to work each morning.

He was realistic about the long-range future: "I'm not planning so much anymore. I will decide step by step. I can't plan for the long run or for the next five years. It's better to do it that way."

His face lit up in a smile when he recalled October 7 for that was the day that he bought a dog, a Dalmatian, as he had promised himself if he felt well enough to handle the responsibility. That was an important step. He knew that under interferon he could not have looked after a dog.

Gleevec had brought him a dog—and life looked much better.

Judy Orem

Breakfast, Then Six Pills

When Judy Orem, one of the first patients in the Phase I Gleevec trials, began taking STI571 early in 1999, she was 100 percent Philadelphia chromosome positive. By September of that year she was only 5 percent positive.

Then a disturbing turn of events occurred. Her Philadelphia chromosome percentages rose to 25, but at least the answer was apparent: she had been off of STI571 for seven weeks because her white cell count had dropped too low. Still, investigators grew concerned that she might have developed resistance to the drug.

Doctors increased her dosage to 600 milligrams.

Her Philadelphia chromosome percentages dropped after that; and by May 2001 she was 100 percent Philadelphia chromosome negative.

"It was just hard to believe. It took a while to sink in. I said: gee, now I have to plan that I'm going to live. That's when we thought about moving from an apartment to a house."

When we visited Judy in October 2001 at her home in Portland, Oregon, she was still taking 600 milligrams of Gleevec each morning—six pills of 100 milligrams each. Her routine was to have breakfast first and then take the pills, and then carry on normally with the rest of her day. As for side effects, she had some water retention around her ankles; a little puffiness around the eyes; and some weakness in her legs. But she had no serious physical complaints.

She was still sending out her newsletter monthly to 240 people though she was no longer the leader of the study group. She gave that up in December 2000; the leukemia case manager at the OHSU took on that task.

As for her response to Gleevec, she felt "excellent." She had just a little anemia.

She felt excited to have been part of the original Gleevec group, forming the friendships, being on the cutting edge, watching other patients get better. She felt a sense of excitement for those who did not get better as well as for those who showed signs of improving for a while before their final setback.

She remembered one woman in particular; her name was Darlene, but not the previously mentioned Darlene Vaughan. In the early summer of 1999, when blast phase patients were admitted to the Phase I patient trials, Darlene had arrived at OHSU by ambulance after being flown in to Portland. She had failed all other treatment. That was of course typical of blast phase patients. She was told she had a week to live. She went on Gleevec and spent the next three months living a rich, full life, seeing her grandchildren, visiting the zoo, the beach, and the mountains. "She had three months of having so much energy."

Judy recalled how exciting it was to see all of these people.

"That kind of got this whole thing heated up on the Internet. Finding people in the blast or accelerated phases who were beginning to do well on Gleevec was certainly an incentive to all of us to move this drug along as quickly as possible. That was probably more exciting than our just getting our white cell counts under control; because that really showed what this drug could do."

Glossary

Ara-C   One of the older chemotherapy drugs; a clear, colorless liquid given intravenously; most commonly used in the treatment of Acute Myeloid Leukemia, Chronic Myeloid Leukemia, Acute Lymphoid Leukemia, and Lymphomas; normally given on a daily basis for five to seven days.

Bcr-Abl   A kinase is a specific type of enzyme. Enzymes are proteins that help in catalyzing or fueling chemical reactions in the cell. A tyrosine kinase is an enzyme that catalyzes in a specific way: it phosphorylates—transferring phosphate groups to specific amino acids on a protein. With its tyrosines in the phosphorylated form, the protein changes shape and this allows that protein to interact with the next protein in the cascade, like in a jigsaw puzzle. In a normal cell, the phosphorylation of key proteins by a kinase only occurs when there is a specific need to fulfill a function, such as cell proliferation, or cell migration, or even the triggering of a "suicidal" death (called *apoptosis*) to ensure the safe removal of very old and damaged cells. In some cancers, such as CML, the chromosomal translocation disrupts one gene that codes for a kinase (e.g., Abl). Separated from its own "head" and now under the control of another gene's "head" (Bcr) in the hybrid oncogene (Bcr-Abl), the kinase becomes totally deregulated and remains "switched on" all the time. The Bcr-Abl rearrangement is a dangerous partnership because sequences on one segment of Bcr are able to drastically disrupt the control of the kinase. It's as if you had a monster that could turn on a switch in its own body and keep it on all the time. That switch tells the kinase to stay continuously on, thus sending signals to the cell to proliferate even if there is no need for additional cells, even if the cell is not mature; and so the cancerous cell moves prematurely from the bone marrow to the blood, and lives longer than it should. In short, the tyrosine kinase known as Bcr-Abl alters the normal genetic instructions of the cell, jamming the signal that orders the body to stop producing white blood cells.

The result: instead of someone falling in the healthy or normal range of 4,000 to 10,000 white blood cells per cubic millimeter, the same volume of blood in a CML patient has 10 to 25 times that figure. It is the huge amount of white blood cells that is characteristic of CML.

In 1986 and 1987, David Baltimore's research team published two articles in *Science Magazine* that pegged the Bcr-Abl protein as a tyrosine kinase, a type of enzyme that plays a critical role in regulating cell growth and division.

Until then, it was not known that the Philadelphia chromosome resulted in the activation of a tyrosine kinase. Now it had become apparent that a certain tyrosine kinase caused CML.

In effect, Drs. Baltimore and Owen Witte had now identified Bcr-Abl's cancer-causing properties.

BILIRUBIN A measure of liver function and decay of red globules. One of two bile pigments, which are colored compounds, or breakdown products of the blood pigment hemoglobin that are excreted in bile.

BLAST CELLS AND CML A CML patient moves into the accelerated phase with 15 percent blast cells and into the blast (final) phase with over 30 percent. A CML patient can be 100 percent Philadelphia chromosome positive without the patient having too many blast cells in either the bone marrow or the blood. But if there is a steady trend upward in either blood or bone marrow in that patient, that is a red flag that the disease may be progressing. In practice virtually 100 percent of the blasts will be Philadelphia chromosome positive (if examined by using molecular techniques).

BONE MARROW TRANSPLANT   Traditionally the one cure for CML; but the procedure is highly dangerous, and extremely high-risk: 70 percent of the patients who undergo this procedure do not survive beyond a year. In the procedure, cancerous bone marrow is killed with high doses of chemotherapy and radiation and then replaced with healthy bone marrow.

CELL   The smallest living component of the human organism able to form energy, comprises billions of molecules. Molecules themselves are combinations of atoms.

CHEMOTHERAPY   Using chemical substances to treat or prevent a disease.

C-KIT   An enzyme; a positive report for c-Kit confirms that a patient has GIST (see GIST). Scientists have learned that c-Kit positive GIST patients generally have a mutation in one of several exons on their c-Kit gene. They have also learned that the response to Gleevec varies according to which exon is mutated. The most common one is exon 11 and that is the one that responds best to Gleevec.

CHRONIC MYELOID LEUKEMIA   Known routinely as CML. Myeloid means of or related to the marrow; a hematologic stem cell disorder caused by an acquired abnormality in the DNA of the stem cells in bone marrow. The first of CML's three phases is the chronic phase, lasting from three to four years. Two-thirds of the patients move to the accelerated phrase that lasts from three to nine months, and eventually to a final phase (known also as the blast crisis), that lasts from three to six months. Each phase is marked by a progressive increase in the amount of white blood cells. It is one of the very first illnesses for which a unique chromosomal abnormality was discovered (known as the Philadelphia chromosome). It is also one of the first cancers that responded very well to interferon. Some of the most impressive results of bone marrow transplantation come from patients with CML. CML always involves the blood and bone marrow, as well as the spleen and liver; twice as frequent in men as in women; more commonly seen after age 50 but can be seen in any age group, even in children.

CHROMOSOME   One of the threadlike structures in a cell nucleus that carry the genetic information in the form of genes; composed of a long double filament DNA coiled into a helix together with associated proteins, with the genes arranged in a linear manner along its length. The nucleus of each human cell contains 46 chromosomes, large-sized molecules, 23 of maternal origin, 23 of paternal origin; each chromosome is comprised of numerous genes that encode the sequences that are the blueprints for proteins.

DNA   The basic genetic material of the cell and the building block of all kinds of proteins. Located in the nucleus of the cell, arranged in 46 chromosomes. It was only in the mid-20th century that scientists began inquiry into the material of which genes were made. Experiments in the 1940s and early 1950s showed that the chemical transmitter of genetic information was DNA, or deoxyribonucleic acid. The major turning point in unraveling of DNA occurred on February 28, 1953 when English physicist Francis Crick and American biochemist James Watson came up with the structure of deoxyribonucleic acid. They found that genes were made up of four chemical units that were arranged along each of two complementary strands; the units form a code of instructions required for an organism's growth and reproduction. The Watson-Crick discovery would eventually set certain scientists on a path that led them to focus cancer research on molecular genetics.

EGF   Epidermal growth factor. See growth factor.

ENZYMES   Proteins that help in catalyzing or fueling chemical reactions in the cell.

EXON   An exon is a segment of DNA within a gene that codes for a protein, separated from adjacent exons by noncoding segments called introns.

FISH (FLUORESCENT *IN SITU* HYBRIDIZATION)   A test for the analysis of chromosomes and DNA by use of fluorescent DNA probes and microscopy that test only 200 cells.

GENE   The basic unit of genetic material, which is carried at a particular place on a chromosome. We each carry a set of genes that determine our individual height, eye color, and numerous other traits; we also harbor genes that can give us cancer.

GLIOBLASTOMA   A very aggressive brain cancer.

GIST   For some time, cancer researchers have been studying the genetic mutations involving the Gastrointestinal Stromal Tumor (GIST). Like CML, it is a rare disease. About 2,000 Americans develop this cancer every year. Patients with inoperable tumors die within two years. The c-Kit enzyme, a defective version of a tyrosine kinase enzyme, exists in GIST.

GLEEVEC   In CML, a defect in the genetic material in the stem cells of the bone marrow causes white blood cell overproduction. When bits of DNA are swapped between chromosomes 9 and 22, the result is the so-called Philadelphia chromosome, whose defective gene produces the protein Bcr-Abl. Disrupting the bone marrow's normally well-controlled production of white blood cells, this abnormal protein (Bcr-Abl) sends continuous signals throughout the cell, leading to a huge increase in the number of white blood cells. Gleevec blocks these signals, potentially preventing the abnormal growth and production of CML cells.

GROWTH FACTOR   Any of various chemicals, particularly polypeptides, that have a variety of important roles in the stimulation of new cell growth and cell maintenance. They bind to the cell surface on receptors. Specific growth factors can cause new cell proliferation, for example, the epidermal growth factor (EGF); and play a role in wound healing (platelet-derived growth factor—PDGF.

HYDROXYUREA   Oral form of chemotherapy that controls the white blood count but does not get rid of cancer cells. Patients can live three to five years on the drug. Often given to lower white blood count in order to prepare for more definitive therapy. Starting definitive therapy with too many white cells in the blood and bone marrow can cause side effects.

INTERFERON   A glycoprotein substance that is produced by cells infected with a virus and has the ability to inhibit viral growth. When interferon burst on the medical scene in the early 1980s, it was touted as a cure for various cancers. Even before patient trials had begun on the drug, the medical community was abuzz with reports that interferon was cancer's holy grail. Alas, it was no holy grail. As it turned out, interferon did prove effective in prolonging the lives of certain cancer patients, but it fell far short of a cure. Some patients were living as long as ten years under interferon treatment.

KINASE   A specific type of enzyme. An agent that can convert the inactive form of an enzyme to the active form. Also: an enzyme that catalyzes the transfer of phosphate groups.

LEIOMYOSARCOMA   The Stage 3 leiomyosarcoma that is mentioned in the book is a rare cancer that occurs about four times in a million, and is very resistant to traditional therapies including radiation, chemotherapies and further surgeries.

LEUKEMIA   A cancer of the blood, bone marrow and liver characterized by an abnormally high rate of white cell production.

ONCOGENES Essentially damaged cells that contain genetic instructions that are flawed and can cause the cell to become cancerous. While normal cells divide, replicate, and die off millions of times during the course of someone's life, during that process, small mistakes occur that are built into the oncogenes. Those mistakes can take decades before the genes go crazy, releasing a spurt of growth signals that order the cell to begin dividing and spreading very quickly.

LEUKAPHERESIS A process to remove white cells in which the patient is attached to a machine and white cells are removed and red cells returned to the patient. This lowers the white cell count within hours and the cells can be stored (frozen) and used later for an autograft procedure.

PDGF-R Platelet-derived growth factor. See growth factor.

PEG-INTERFERON A form of interferon called pegyllated interferon or PEG-interferon. It offers reduced side effects (compared to the standard interferon-alpha that CML patients are routinely given) and once-a-week injections (rather than daily ones). It appeared to be, at the time, more effective than interferon-alpha in treating leukemia but a recent study has thrown this into question. It has been estimated that at diagnosis the average patient has between 100 billion and one trillion leukemic cells in the body. Since resistance of cancer cells to a drug is based on probability—the more cells, the greater the chances of resistance—one might reduce the chances of resistance by reducing the number of leukemia cells in the body.

The "peg" of pegyllated stood for polyethylene glycol (the same as antifreeze!), a chemical that helped the interferon circulate in the body longer, providing more continuous exposure of the cells that are to be killed to the drug. The reason: the half-life of interferon-alpha in the body is a few hours; of PEG interferon, one week.

PHILADELPHIA CHROMOSOME   When bits of DNA are swapped between chromosomes 9 and 22, the result is the so-called Philadelphia chromosome, whose defective gene produces the protein Bcr-Abl. Although no one knows what causes this DNA change, it is clear that the resulting abnormal protein disrupts the bone marrow's normally well-controlled production of white blood cells. This "deregulated" production of white blood cells leads to a massive increase in their concentration in the blood, an indication of CML. The Philadelphia chromosome was the first chromosomal defect linked to a cancer.

PCR (POLYMERASE CHAIN REACTION)   A test that checks 100,000 cells; far more comprehensive than the FISH test (see above).

SIGNAL TRANSDUCTION INHIBITORS   These drugs—and Gleevec is one of them—are so called because they interfere with the pathways that signal the growth of cancerous cells. When an STI interrupts a signal transduction pathway, a cell stops dividing, halting the cancer. The exact process in signal transduction works like this: a growth-factor protein binds to a receptor at the outside of a cell. Various signaling enzymes then transmit that signal to the cell nucleus where it can eventually trigger cell growth.

TYROSINE KINASES   Enzymes that activate the signaling proteins by adding phosphate groups to them; an enzyme that catalyzes in a specific way: it phosphorylates—transferring phosphate groups to specific amino acids on a protein. A small compound lodged in the right place can often close an enzyme down.

WHITE BLOOD CELL COUNT   An increased number of white blood cells tends to confirm a diagnosis of CML. The normal range is between 4,000 and 10,000 per cubic millimeter.

Acknowledgments

We needed a great deal of help in telling the story of Gleevec.

When we began the project in April 2001, neither of us had a sense of how easy or difficult it would be to recapture the main highlights of the drug's discovery and development.

We had no idea, for instance, how willing patients taking Gleevec would be to review their case histories with us. To our pleasant surprise, we found almost every patient to whom we turned willing to share their brave and lengthy struggles with their disease; and eager to share those shining moments when Gleevec entered their lives.

We also had little notion of what parts of the story needed telling. So many people took part in the discovery and development of Gleevec over so many years; literally thousands of people at Novartis alone had some role in the discovery and development of Gleevec.

By talking to countless people, we began to develop an idea of what to include in the book.

We wish that it were possible to thank each person who made a contribution to the discovery, development, and distribution of Gleevec. Likewise we wish that we could express gratitude to everyone who helped us in the researching and writing of this book.

In both cases, it is not possible.

Nevertheless we wish to thank a number of people who made major contributions in helping us craft the book.

Critical to so many aspects of the Gleevec story is Alex Matter. He found time to talk with us at length about the history of Gleevec's discovery and development; he also helped to make sure that we portrayed the scientific and medical aspects of the story as accurately as possible. Alex, a huge thank you from both of us.

Two other key contributors to this book were Jürg Reinhardt and Andreas Rummelt, both senior executives at Novartis. It was Jürg and Andreas who made sure that the plan to accelerate the development of Gleevec was implemented; both men spent evenings, weekends and vacation time to move the project rapidly ahead. We thank them both for reliving for us their part of the Gleevec story.

We were particularly pleased to have a chance to talk with such key players as Jürg Zimmermann, Elisabeth Buchdunger, and Brian Druker. These were the scientists among several others who literally brought Gleevec to life, in the case of Zimmermann and Buchdunger; and who saw it through its successful journey in the early patient trials, in the case of Brian Druker.

A real highlight of the research for this book was the series of conversations we had with Gleevec patients. Those conversations were deeply moving for us. We could not help but be impressed with their courage, their intelligence, and their perseverance. All of these patients took our project with the utmost seriousness, spending hours with us to make sure we understood their remarkable battles. Never once did they give us the impression that we were invading their privacies. We admire them all and are most grateful to all of them, wishing them only the best of health.

Given the constraints of space in a book like this, we were not able to tell each story in full; some are not told at all. But every patient to whom we talked helped us to convey the story in all of its breadth. And so we thank each patient for being a part of our book project.

Those with whom we spent time and listened to their stories are: Roger Baker; Sandra Craine; Nigel Douch; Nora Flanzbaum Friedenbach; Suzan McNamara; Marco Nese; Judy Orem; Keith Pratt; Eliza-

Acknowledgments

beth Rees; Richard Rockefeller; Anita Scherzer, and Darlene Vaughan. We also want to say thanks to Ophir Zelinger, who when his father-in-law Ehud Nehemya was too ill to talk with us, filled in a very detailed portrait of this very brave man.

Many people helped to round out the story and we wish to thank them as well: Lorraine Armstrong; Shimon Binyamini; Kathy Bloomgarden; Paula Boultbee; Jurgen Brokatzky-Geiger; Gregory Burke; Renaud Capdeville; Jerh Collins; Odette Cunico; Burkhard Daldrup; Robert Doyle; Deborah Dunsire; Jim Elkin; David R. Epstein; Martin Fey; John M. Ford; John M. Goldman; Peter Graf; Sonja Herr; Lukas Jauslin; Max Kaufman; Barbara Kennedy; Alexander Levitzki; Martha Lichtensteiger; Junia V. Melo; Robert Miranda; Ulrike Pfaar; Peter Rowbotham; Beat Ruttimann; Charles L. Sawyers; Norman J. Scherzer; Gloria C. Stone; Franz Sutter; and Elinore Y. White.

To our editors, David Conti and Edwin Tan, we offer our gratitude for showing such care and attention to the project; both made supreme efforts to turn our own efforts into a book that was easier and simpler to read. It was a pleasure working with both of you. Thanks.

Dr. Daniel Vasella, M.D.
Robert Slater
December 2002

About Novartis

Novartis AG, a Swiss-based holding company, is a world leader in the fields of pharmaceuticals and consumer health. In 2002, the group's businesses achieved sales of CHF 32.4 billion (USD 20.9 billion) and a net income of CHF 7.3 billion (USD 4.7 billion). The group invested approximately CHF 4.3 billion (USD 2.8 billion) in R&D. Headquartered in Basel, Switzerland, Novartis Group companies employ more than 74,000 people and operate in more than 140 countries around the world.

Daniel Vasella is the chairman and CEO of Novartis AG. Dr. Vasella is also the chairman of the board of directors of Novartis Corporation, the group's U.S. holding company and Novartis Pharma AG, the group company that leads the global pharmaceuticals business. In this latter role, Dr. Vasella provides leadership, support, and guidance to the group's pharmaceuticals business, especially where substantial resources are to be expended to create products such as Gleevec.

You have found in this book the words "we," "our," "us," "company," and similar words or phrases. Use of these words should not be taken to mean that any one affiliate exercises dominion or control over another affiliate. Indeed, that is not the case as each operating company in the group is legally separate from all other companies in the group and manages its business independently through its respective board of directors or other top local management body. No group company operates the business of another group company nor is any group company the agent of any other group company. The language used in this book is meant to convey the close coordination between management, scientists, and other associates of interdependent but legally independent companies that is necessary to discover, develop, and bring to market exceptional pharmaceutical products in today's environment.

For further information about Novartis, please consult http://www.novartis.com.

About Novartis

Index